TAHOMA
and its People

TAHOMA
and its People

A Natural History of Mount Rainier National Park

JEFF ANTONELIS-LAPP

WSU
PRESS

Washington State University Press
Pullman, Washington

WSU PRESS
WASHINGTON STATE UNIVERSITY

Washington State University Press
PO Box 645910
Pullman, Washington 99164-5910
Phone: 800-354-7360
Email: wsupress@wsu.edu
Website: wsupress.wsu.edu

Library of Congress Cataloging-in-Publication Data

Names: Antonelis-Lapp, Jeff, 1954- author.
Title: Tahoma and its people : a natural history of Mount Rainier National
 Park / Jeff Antonelis-Lapp.
Description: Pullman, Washington : Washington State University Press,
 [2020] | Includes bibliographical references and index.
Identifiers: LCCN 2019041253 | ISBN 9780874223736 (paperback)
Subjects: LCSH: Natural history--Washington (State)--Mount Rainier National
 Park. | Indians of North America--Washington (State)--Mount Rainier
 National Park. | Natural history--Washington (State)--Rainier, Mount. |
 Mount Rainier National Park (Wash.)
Classification: LCC QH105.W2 A58 2020 | DDC 508.797--dc23
LC record available at https://lccn.loc.gov/2019041253

Cover photo by Nat Chittamai/Shutterstock.com

To the Silver Hummingbird, a true believer

Before the world changed, five sisters lived near Orting. When Doque-buth, the Changer, came, he changed them into five mountains. One of them he called Takkobad—Mount Rainier. Doquebuth said to Takkobad, "You will take care of the Sound country. You will supply water. You will be useful in that way."

Jerry Meeker, Puyallup Tribe
From Ella E. Clark, *Indian Legends of the Pacific Northwest*

Contents

Introduction

A postcard day of exquisite beauty and splendor greets me, a lone traveler on the mountain's northeast slopes. Thin bands of wispy cirrus clouds laze across the azure sky. An eerie, windless quiet prevails as I follow a faint boot track in the melting snow to Frozen Lake. Lemon-colored glacier lilies nod their greeting, signaling high summer. I pause to marvel, mop my brow, and catch my breath. When I remove my sunglasses, the sun's blinding glare sears my retinas.

The road to Sunrise remains closed today because of helicopter operations in the area. I appear to be the only person on the trail leading into the snow-laden, subalpine meadows of Yakima Park. The mountain's icy upper slopes recently claimed another life, and climbing rangers hope to remove the body from a location high on the Winthrop Glacier.

I watch transfixed as two choppers shuttle equipment and personnel onto the glacier. The staccato *whump-whump-whump* pounds out a grim reminder that this mountain can be treacherous and unforgiving. When the hulking Chinook helicopter finally lifts off for the last time, its joyless task complete, it thunders directly overhead, barely above the treetops. I stop reflexively, doff my cap, and bow my head.

It never occurred to me when I came to the mountain today that I might watch a rescue team recover a body from a climbing accident. I came for the same reasons that usually drive me: To spot my first black bear of the season, rooting about a rotting log in search of its next meal. To watch a glacier give birth to a wild river that races and rumbles off to the lowlands. To linger in meadows adorned in a kaleidoscopic pattern of wildflowers. To visit small camps where people lived seasonally, hunting hoary marmots and mountain goats, gathering huckleberries and bear grass.

I came to continue my practice of natural history, the study of living things in their natural setting. This book explores the interrelationships within and between plant and animal communities, the influence that weather and climate exert on ecosystems, the geologic processes that create and alter landscapes, and the relationship between the iconic mountain and the people who have traveled to it for millennia.

A ranger once told me that Mount Rainier is one of the world's largest outdoor science labs. My contact with the many scientists and technicians working on and around the mountain confirms this comment. Their willingness to discuss their research and share field time added an irreplaceable dimension to my work. They took me onto the Nisqually Glacier to help measure ice velocities, a possible warning sign of life-threatening glacial outburst floods. I watched 30 firefighters contain a 60-acre prescribed prairie burn, and then returned the next spring to appreciate the bountiful wildflowers that benefit from periodic fires. Experts taught me to identify stone tool fragments at archaeological excavations where the ground held untold stories of human presence. Community officials invited me to observe an evacuation drill for 2,000 students and local residents who hiked nearly two miles to a safety zone in preparation for a devastating debris flow.

Mount Rainier National Park contains some of the most protected landscapes on Earth. Designated as our fifth national park in 1899, it was the first to have a master plan as the guiding document to restrict development and permanently protect park resources. The Washington Park Wilderness Act of 1988, sponsored by Senator Dan Evans and signed into law by President Ronald Reagan, added another layer of preservation by declaring 97 percent of the park as wilderness. The remaining 3 percent that includes buildings and roads attained protected status as a National Historic Landmark District in 1997.

Despite these safeguards, the grand mountain that towers 7,000 feet over its neighboring peaks is under siege. Its cubic mile of ice and snow—enough to fill a string of railroad tanker cars stretching to the moon—has shrunk by nearly one-fourth in just 30 years. This dramatic reduction foreshadows myriad changes for the region. One that concerns park staff, scientists, and municipal officials is the effect of Mount Rainier's receding glaciers on its rivers and in Puget Sound. Having retreated further upslope than at any other time in recorded history, the glaciers release great quantities of sediment and other debris into their rivers. Known as aggradation, this accumulation fills river bottoms with sediment. As they fill, the channels hold less water during high volume events, spilling from their banks more readily to increase the risk of flood damage. The sediment eventually makes its way into the Sound at a rate of about one million tons per year.

TERMINOLOGY

Throughout the book, I use a variety of names that warrant clarification. Beginning with the mountain itself, British explorer Captain George Vancouver upheld an abiding western European tradition when he named the distant, snow-covered peak for Admiral Peter Rainier in 1792. The naming of landforms and other geographic features served multiple functions, and doing so in honor of others signaled a gesture of fondness, gratitude, or respect. Had Vancouver known that the people living for generations within sight of the volcano already called it by name, it would have wielded no influence whatsoever on his naming decision.

In his 1963 interviews of tribal elders living near the mountain, Washington State University anthropologist Allan H. Smith learned that the Yakama people referred to it as *Taxóˑma*. He reported that the Muckleshoot knew it as *Taqóˑbid* and the Nisqually as *Taqóˑma*. Nisqually tribal historian Cecelia Svinth Carpenter recorded the mountain's name as *Ta-co-bet* or *Tacobet*. Other names include linguistic variations of *Tahoma* or *Takhoma*, all thought to have been used by various tribal peoples. There is no universal agreement on a single meaning. Elders, linguists, and other students of the region's indigenous languages place the meaning somewhere between "the source of all waters," "white mountain," or "snow peak." Some believe that *Tahoma* described all snow-capped mountains.

I use Mount Rainier and Tahoma interchangeably, especially employing the latter in relationship to Native Americans. Much of the literature and many local residents call it "the Mountain," a term I also use, without capitalization.

Puget Sound lies less than 50 miles north and west of Mount Rainier and drains nearly all of its rivers. The Sound extends roughly 100 miles north from Olympia along a north-south axis. To recognize the larger ecosystem to which it belongs, Canada and the United States adopted "Salish Sea" as the official name for the inland waters that stretch from Washington into British Columbia. With a coastline that approximates one-third of the total of the contiguous United States and Alaska coasts, this jewel glimmers as one of the world's most magnificent inland seas. Salish Sea honors the Coast Salish people who plied its waters and fashioned complex societies along its shores for eons prior to the arrival of Russian and European explorers. Native American communities and tribal nations today continue to demonstrate leadership across boundaries of politics, culture, and natural resources. Puget Sound generally prevails as the term of common usage among area residents, but I often use Salish Sea, particularly in reference to native people.

To name those who first fished the Salish Sea and its tributaries, and harvested its shellfish bounty from exposed tideflats, I take my cue from the Native American students and colleagues with whom I worked, taught, and learned for over 20 years. Borrowing the words that they use to describe themselves, I interchangeably use Native American, first people, indigenous, and native for the first residents who by at least 14,000 years ago began settling the region after the Puget Lobe of the Cordilleran Ice Sheet receded. Some readers may consider Indian inappropriate, but its use by Native Americans in the region suggests otherwise. Readers will also note that *Yakama* refers to the indigenous people of central Washington, while *Yakima* denotes the city and valley of that area.

A scientific name identifies an organism taxonomically by its genus and species, for example, *Homo* (genus) and *sapiens* (species). Scientific names are extremely important in botany where a proliferation of aliases for the same plant produces considerable confusion. I therefore provide scientific names for species possessing several common names but otherwise use them sparingly. I also explain the Greek or Latin roots and the meaning of those terms that I find interesting or that supply additional information about the organism.

Speaking of plants, the text frequently discusses their use by first peoples as food, medicine, or for other purposes. Park regulations strictly prohibit plant collecting and experts should supervise the preparation and use of plants for any purpose.

ORGANIZATION

This book begins with the geologic history of the mountain, followed by an introduction to the Native Americans who have lived around and visited Tahoma for over 9,000 years. The remaining chapters progress clockwise around the mountain, the direction most people hiking the 93-mile Wonderland Trail take as they encircle the mountain. Chapter three follows the Nisqually River watershed from the terminus of the Nisqually Glacier to its runout into Puget Sound. Chapter four visits Longmire, the site of the original park headquarters. Chapter five traces the Puyallup River from the glacier to Tacoma's Commencement Bay. Chapter six follows the Carbon River, and an exploration of the Sunrise area completes the trip around the mountain.

When giving presentations on Mount Rainier, I sometimes distribute index cards and ask people to write about their most treasured memory of the mountain. I then collect the cards and read them aloud. Some of my favorites include: "My family and I went camping there every year when we were kids and it was the highlight of our summer vacation." "My husband proposed to me at Mount Rainier." "My grandparents told me stories of *their* grandparents going to the mountain every summer to pick huckleberries."

These intimate vignettes, shared among strangers, exemplify the deep attachment that an untold number of people feel to Mount Rainier. It seems that nearly everyone who visits, works, or lives near the mountain develops a special kinship with it, a set of stories about it. This book is for them, and many others. It is for those whose personal bonds span decades, for families who have created generations of memories, and those just beginning to collect them. It is for the million-plus annual visitors, some of whom travel great distances while others treat it as their own backyard, and for the native people living around it whose ancestors breathed life into the stories we continue to learn. *Tahoma and Its People* is for anyone who loves this incomparable treasure and wishes to know it better.

1

Mount Rainier: Geologic History and Processes

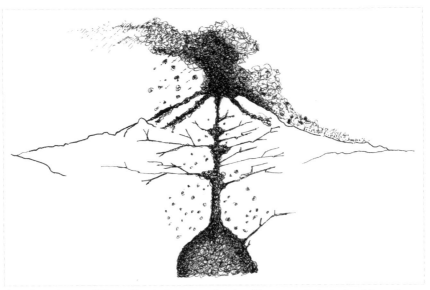

Artist's rendering of Mount Rainier's magma chamber. The mountain is depicted from the east; magma chamber based on information in John et al. (2008). *Drawing by Lucia Harrison*

ANCESTOR MOUNTAINS

Mount Rainier, a looming volcano, has attracted people for millennia. Most of its visitors, untrained in reading its landscape stories, never think about the mountain's changing geological faces. Geologists and other landscape readers, however, learn to ask questions, sleuth for answers, and discover environmental clues that unlock stories held in rock formations and sediment layers. These experts reconstruct geologic events that guide our understanding of land processes, define research pathways, and inform land management and disaster preparedness plans.

No one was more adept at interpreting Mount Rainier's geology than the small group that pioneered landmark work there in the 1950s and 1960s. Among them were U.S. Geological Survey geologists Donal R. Mullineaux and Dwight R. "Rocky" Crandell, who led the way in uncovering much of the foundational knowledge about the mountain's volcanic history. One look at Crandell's 1953 field notebook verifies his knack for piecing together stories from clues in the landscape. In one journal entry, he sketched a light bulb next to the word "Idea!" to mark the moment he realized that the rock, clay, and sand deposits on the Enumclaw Plateau differed from any he'd seen before. He suspected they were part of something not yet described. Subsequent research confirmed his hunch, charting a new course of study for debris flows at Mount Rainier.

Contemporary Mount Rainier represents just the latest snapshot of uncountable geologic stories told over millions of years. It begins with the heavy and dense Juan de Fuca Plate, a slab of Earth's crust that comprises the ocean floor off our West Coast. As it inches eastward, the plate collides with and dives under another piece of Earth's crust, the lighter and less dense 30-mile-thick North American Plate. These plates are part of the upper layer of the 60-mile thick lithosphere (from the Greek *lithos*, for

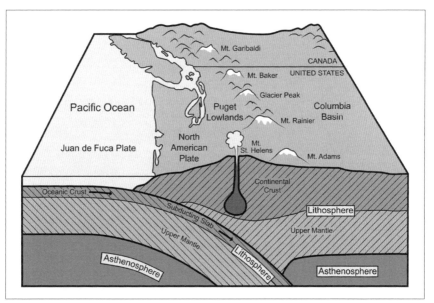

The tectonic plates that undergird the Pacific Northwest create the Cascade Range and a host of related features. *U.S. Geological Survey, redrawn by Kirsten Wahlquist*

stone). They ride upon the highly viscous asthenosphere (from *asthenēs,* for weak). This process of subduction begins around 50 miles offshore, crawling about one inch per year, a rate comparable to human fingernail growth. At this laborious pace, the Juan de Fuca Plate creeps forward about 20 miles every million years. The inexorable movement of the last 40 million years has helped create Mount Rainier and the other geographic features that define the Pacific Northwest.

The tremendous continuing collision causes the Juan de Fuca Plate to sink beneath the North American Plate, dragging sediment and water to a depth of about 60 miles. There, pressure and heat partially melt the rocks under the North American Plate, creating magma. Mostly basalt, this buoyant, molten rock travels paths of least resistance upward toward Earth's surface. It exits through crustal areas weakened by fractures in the parent rock called fault zones. Mount Rainier and most of its tallest neighbors—Mount Baker, Glacier Peak, Mount Adams, Mount St. Helens, and Mount Hood—sit along such fault lines. Gas-rich magma powers the volcanoes that form the Cascade Range volcanic arc that stretches 780 miles from Mount Garibaldi in British Columbia to Mount Shasta and Lassen Peak in northern California. Each consists of alternating strata—or layers—of lava flows and other volcanic ejecta that includes ashfall and tephra (particles larger than sand grains). These are called composite or stratovolcanoes.

Anchoring the southern end of the northern section of the Cascades Range, Mount Rainier displays three vital signs of a living volcano. The first is the occurrence of earthquakes beneath and near it. Eleven seismometers on or near Mount Rainier record three to 10 earthquakes per month, most of them undetectable by anything other than highly sensitive seismographic instruments. Small groups of 10 or more earthquakes also occur occasionally, likely the result of stresses related to hot fluids circulating under the mountain.

Because even a small earthquake can launch a havoc-wreaking debris flow, various agencies keep a close watch on the volcano. Scientists at the U.S. Geological Survey's Cascade Volcano Observatory in Vancouver, Washington, those at the University of Washington's Pacific Northwest Seismic Network in Seattle, and National Park Service geologists at Longmire closely monitor equipment that tracks the mountain's every move. Precision instruments located between 7,000 and 11,000 feet on the mountain detect the slightest disturbances and then immediately transmit the data to scientists. Seismometers measure and record the slightest tremor.

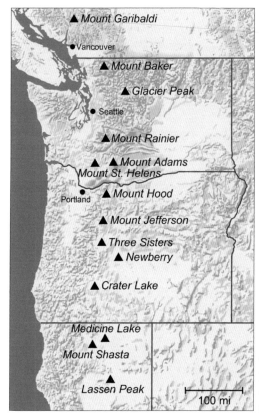

▲ Mount Garibaldi

● Vancouver

▲ Mount Baker

▲ Glacier Peak

● Seattle

▲ Mount Rainier

▲ ▲ Mount Adams
Mount St. Helens

●
Portland   ▲ Mount Hood

▲ Mount Jefferson

▲ Three Sisters

▲ Newberry

▲ Crater Lake

Medicine Lake
▲ ▲
Mount Shasta

▲
Lassen Peak

100 mi

The major peaks in the Cascade Range, spanning from Canada to California. *U.S. Geological Survey, redrawn by Kirsten Wahlquist*

Tilt meters track changes in the mountain's surface to within a half inch. A Global Positioning System (GPS) network relays the exact location of any activity. This array of sensing equipment should help provide advance warning of an event at one of the region's most seismically active volcanoes.

The second sign of an active volcano is the presence of mineral springs. As gases from the mountain's core ascend to the surface, they bubble up through water in the Ohanapecosh and Longmire areas of the park.

Vents on the mountain that emit gases and vapors as hot as 187° Fahrenheit—the boiling point at the summit and 60 degrees hotter than the recommended setting for household hot water—are the third sign that Rainier is a living volcano. Emissions from these vents, called fumaroles, signal geothermal activity in the magma chambers deep within the volcano's interior. The heat melts snow from summit snowstorms, often leaving rocks bare and visible for many miles.

Long before the Cascade volcanic arc pulsed to life about 40 million years ago, Mount Rainier's foundation was already under construction. Between 50 and 60 million years ago, portions of the Pacific coastline lay 60 miles east of the current coast, at approximately the same location as today's Interstate 5. These coastal lowlands consisted mostly of swampy marshland. Sand and clay sediments collected there, brought in by rivers that drained the lands to the east. Plant communities formed on top of the mineral-rich sediments. An alternating pattern developed where fresh deposits of river sediments covered the plants. Over time, the communities re-established themselves. This repeating process resulted in the accumulation of sand, clay, and plant materials that over millions of years compacted to form shale, sandstone, and coal. Today's layers of sedimentary rocks, nearly two miles thick and 34 to 48 million years old, include rocks of the Puget Group.

The first of the Cascade Range volcanoes fired up across this broad coastal plain, erupting between 37 and 43 million years ago. The North-craft Formation, one of the early Cascade volcanic centers, lies about 35 miles southwest of Mount Rainier.

Beginning around 37 million years ago, a series of eruptions sent mighty mudflows and churning rivers of volcanic debris both eastward and westward across the lowlands into the Pacific Ocean. The main points of origin centered around Mount Wow, about eight miles southwest of the mountain's summit, and in the Cowlitz Chimneys area, about seven miles east. These eruptions brought the first topographic relief to the area, creating a landscape of hills and valleys up to several thousand feet in elevation at Mount Rainier's present location. Active for 10 million years and known as the Ohanapecosh Formation, it was the first of three ancestral eruptive periods in the Mount Rainier area. Visitors entering the southwest corner of the park at the Nisqually Entrance drive on its remnants.

Hot and mobile, high volume flows of pumice or pyroclastics ranging to several hundred feet thick marked the second ancestral eruptive period. Pumice, a light and porous rock, forms when frothy lava cools and solidifies quickly, trapping the air bubbles inside. Pyroclastic flows, deadly clouds of super-hot gases and rock fragments, race downslope at ground level. Most flows originated outside the present park boundaries, coursing down the sides of multiple volcanoes to cover vast areas. Preserved as the Stevens Ridge Formation, these flows occurred between 22 and 26 million years ago.

Massive lava flows, viscous and slow moving like steaming asphalt, highlighted a third volcanic ancestral period contemporaneous with Stevens Ridge. Highly explosive volcanoes boiled at multiple sites, producing and expelling so much magma that they collapsed and formed calderas. Lava from one flowed to between 2,500 and 5,000 feet thick. The Mount Aix caldera produced about a hundred times the volume as the 1980 Mount St. Helens eruption. These and others represent the Fifes Peak Formation, active between 22 and 27 million years ago.

Although the Ohanapecosh, Stevens Ridge, and Fifes Peak periods are the principal volcanic formations upon which Mount Rainier sits today, other events and processes contributed to its geologic beginnings. The

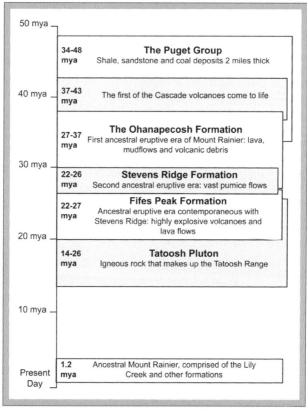

Geologic precursors of the modern-day Mount Rainier volcano (mya=millions of years ago). *Kirsten Wahlquist*

Tatoosh Pluton, a mass of igneous rock that formed and cooled under Earth's surface, spans much of the Tatoosh Range along the park's southern border. Between 14 and 26 million years old, the rock also appears on the mountain's north side in the Sunrise area.

Closer to present times, an ancestral Mount Rainier rose from multiple volcanic eruptions between one and two million years ago. Although very little remains of it, the 1.2 million year-old Lily Creek Formation formed from a volcano that was probably similar to today's mountain. It featured a series of high ridges and deep valleys that radiated from a central summit.

This series of geologic events, beginning with and building upon the Puget Group, constitutes the primary antecedents of present-day Mount Rainier. The Ohanapecosh mudflows, the Stevens Ridge pumice and pyroclastic flows, and the Fifes Peak explosive volcanoes with their enormous lava flows all figure as foundational geology for the Mount Rainier area. The magmatic activity that produced the Tatoosh Pluton and Lily Creek Formation also made its mark, setting the stage for the creation of the solitary glacier-covered peak that dominates the southeastern skyline of the Salish Sea region.

CREATIVE FORCES, DESTRUCTIVE POWERS

Ample evidence suggests that construction of the present Mount Rainier volcano began about 500,000 years ago. To appreciate its geologic history, it's helpful to imagine key events occurring within a 24-hour day. On this clock, one hour represents approximately 21,000 years; one minute about 350. An 80-year lifetime would skip past in less than 15 seconds. Mountain building began at 12:00:01 a.m., a half million years ago. Today's date corresponds to the end of that day, 11:59:59 p.m.

A long series of explosive volcanic eruptions—some ejecting rock material 15 miles down valley—gave birth to a new Mount Rainier. Several massive lava flows with volumes approaching two cubic miles oozed forth. The flows created Old Desolate, Burroughs Mountain, and Grand Park on the mountain's north side and unnamed ridges in the Mowich River area in the park's northwest section. They all bear evidence of contact with ice and snow during their deposition, suggesting that the adjacent valleys brimmed with ice at the time of the flows. The foundation and frame of today's mountain settled into place during this period, between midnight and about 5:00 a.m., or 500,000 to 420,000 years ago.

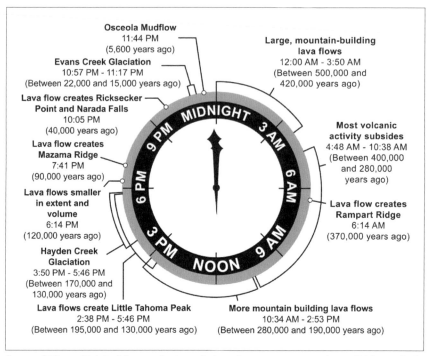

Some of Mount Rainier's key geologic periods and events represented on a 24-hour clock (1 hour=21,000 years). Mountain building began at 12:00:01 a.m.; the present time is 11:59:59 p.m. *Kirsten Wahlquist*

The mountain then entered an interval of relatively little volcanic activity, which later eroded away and left scant evidence for geologists to interpret. Between 4:48 and 10:34 a.m., from 400,000 to 280,000 years ago, fewer and smaller flows covered less ground than earlier events. This probably coincided with substantial erosion on the upper slopes, reducing the mountain's height. One exception to this calmer time was the Rampart Ridge lava flow near present-day Longmire at 6:14 a.m., 370,000 years ago.

A series of mountain-building lava flows began soon after, probably extending the peak to its greatest height. Geologists in the 1960s calculated that the angle of dipping lava beds near the summit indicated that it towered above 16,000 feet, but recent research places the mountain's prior height at closer to 15,000 feet. Lava flowed with a volume so great that magma burst through the upper eastern and western flanks to create gigantic flows through side vents. Some magma remained in the flank cracks to cool and harden. Over the ages, its heat and acidic gases weakened the surrounding rock. Eventually this fatigued, unstable rock would give way to gravity and slide off the mountain to become lahars known as the Osceola, Paradise, Round Pass, and Electron Mudflows, among others. Features built between 10:34 a.m. and 2:53 p.m. include Mount Ruth and Meany Crest in the mountain's northeast sector, Klapatche Ridge and St. Andrews Park on the west side, and Mowich Face and Liberty Cap on the northwest flank. Like earlier flows, they show evidence of contact with ice and snow, indicating that the river valleys contained immense volumes of ice.

Near the end of this period, another lava flow created the base of Little Tahoma Peak. A second and then a third flow completed the rocky crag that stands directly east of Mount Rainier, the state's third highest peak at 11,138 feet.

The time of fire that formed "Little T" with lava flows coincided with a time of ice that transformed Mount Rainier's appearance through a widespread alpine glaciation. Glaciers crept 63 miles down the Cowlitz drainage past Randle to Riffe Lake south of Morton. Named the Hayden Creek glaciation by trailblazing geologist Rocky Crandell for the area in which it was first noted, deposits also appear about 30 miles down the Nisqually valley near the west end of Alder Lake. Lasting roughly from about 3:50 p.m. until 5:46 p.m. (170,000 to 130,000 years ago), glaciers covered most of the mountain's high areas and ground the White, Nisqually, Puyallup, and Mowich drainages into today's familiar U-shaped valleys.

Following the Hayden Creek glaciation, lava flows resumed, creating new landforms. Echo and Observation Rocks rose up on the mountain's northwest flank above Spray Park at 6:14 p.m. At about 7:41, another glacier-bounded flow built Mazama Ridge in the Paradise area. Eruptions and lava flows continued until 11:00 p.m., some 20,000 years ago, building most of the upper mountain. Curtis Ridge, Liberty Ridge, and the nearly vertical Willis Wall all took shape on the north side. On the opposite side, lava inched down and settled between the Nisqually and Paradise Rivers where glacial ice towered hundreds of feet high. When it retreated, it left the 800-foot tall Ricksecker Point lava flow, the only one to travel any distance during this period.

Besides these flows, current dating techniques limit the ways to learn about what happened between 10:00 and 11:44 p.m. Although radiometric dating accurately calculates age by comparing a material's radioactive isotope with its decay products, it doesn't work well on the glassy and young (geologically speaking) rocks of the upper slopes. And with no plants growing that high, radiocarbon dating of plant material isn't possible. Instead, most clues come from studying debris flows and tephra layers, the rock material sent skyward during an eruption that later carpets the ground. Working from these, geologists have found evidence of nearly 40 volcanic eruptions in this period alone.

Just before 11:00 p.m., glaciers once again made their mark on the mountain's high country. They carved bowl-shaped hollows called cirques at the upper end of some valleys. Glacial lakes known as tarns formed in some: Shriner Lake in the east, Lake George at the base of Gobbler's Knob in the southwest, and Mowich Lake in the park's northwest corner. Glaciers blanketed the lowland river valleys, too. They snaked 24 miles down the White River drainage on the north side to within a few miles of Greenwater, down the Cowlitz 38 miles nearly to Randle, and 19 miles down the Nisqually beyond Ashford. Called the Evans Creek glaciation, it peaked by about 11:17 p.m., 15,000 years ago.

Around that same time, other forces sculpted the lowlands west and north of the mountain as giant sheets of continental ice slid southward into present-day Washington. Although not directly connected to Mount Rainier proper, it shaped the basin into which most of the mountain's rivers empty. The ice advanced during colder periods called stades and retreated during warmer spells known as interstades. This pattern continued until the last of at least seven glacial periods sent two lobes of the Cordilleran Ice Sheet from British Columbia's Coast Range into the Puget Sound Lowlands. The Juan de Fuca Lobe spread west along the north side of the Olympic Mountains to the Pacific Ocean. By 11:14 p.m., the monstrous Puget Lobe ran the length of the region, towering nearly one mile thick near Bellingham. It stood 3,500 feet thick in the Seattle area—nearly six Space Needles stacked end to end. It tapered to 2,000 feet in Tacoma and its leading edge reached south of Olympia. The gigantic tongue of ice spanned from the Olympic Mountains to the Cascade Range, and was so heavy that the ground beneath it sunk anywhere from 160 to over 400 feet. It dammed the waterways draining the Cascades, forcing the water south and then west around its terminus. Containing more water than the Columbia River, the flow churned southwest and headed out the Chehalis River valley to the Pacific. Water, one of nature's most powerful and artistic sculptors, flowed under the ice sheet and fashioned much of the Puget Sound area's topography below 3,000 feet. Meltwater gouged troughs in the land that became Hood Canal and Lakes Washington, Union, Sammamish, Tapps, and dozens of others. It also produced the Salish Sea, its most ornate carving.

The maximum extent of the Puget Lobe of the Vashon Stade of the Fraser Glaciation of the Cordilleran Ice Sheet. Note its coverage of the entire Puget Lowlands. Also notice the Chehalis River, the melting ice sheet's main conduit to the Pacific Ocean. *Courtesy of Washington Geological Survey, modified by Kirsten Wahlquist*

A changing climate brought North America's final glacial advance to a halt at about 11:16 p.m. (15,400 years ago). The Puget Lobe retreated northward, broke up, and melted. Its terminus reached the U.S.-Canada border by 11:28 p.m. It disappeared altogether within another three minutes on the clock, signaling the end of the Vashon Stade of the Fraser Glaciation, the region's last Ice Age, around 10,150 years ago.

An explosive lateral blast collapsed the uppermost section of the mountain's northeast side at 11:44 p.m., 5,600 years ago. Within seconds, the unleashed material traveled more than a mile and transformed from an avalanche into a debris flow called a cohesive lahar, one of the largest on record. Exceeding 40 miles per hour, it lurched down the mountainside with so much power and volume that it overran the White and West Fork White River channels. It inundated and incorporated whole sections of forest and undercut banks and slopes until its immense bulk stood equivalent to the height of a 50-story building. The volume was so prodigious that in one slot gorge it created two waterfall-like features that would have dwarfed the largest at Niagara Falls by a factor of six.

Flowing with the consistency of wet cement, the lahar surged out of the foothills and onto the lowlands. It steamrolled the Enumclaw Plateau at an average depth of 60 feet and forged a new channel across it. Sweeping through the present-day cities of Auburn and Kent into Puget Sound and west through Buckley, Sumner, Orting, and Puyallup at about 20 feet deep, it ran out into Tacoma's Commencement Bay. This sudden, violent outburst filled in 60 square miles of the Salish Sea and covered the equivalent of Seattle's land area two and a half times. Its volume of nearly one cubic mile exceeded the 1980 Mount St. Helens flows by 20 percent. These were the deposits recorded in Rocky Crandell's 1953 notebook, later called the Osceola Mudflow.

Numerous volcanic eruptions, lahars down every major river valley, and lava flows that rebuilt the mountain's summit marked the next series of events between 11:52 and 11:54 p.m. Eruptions caused the Round Pass Mudflow, a lahar similar to but smaller than the Osceola, in the park's west and southwest sector. The mountain's upper west slopes gave way and sent debris the height of a 100-story building hurtling downriver that reached as far as Ashford.

Most eruptions from this era—called the Summerland Period—produced such small amounts of volcanic ash that geologists struggle to tell them apart, but the one at about 2,200 years ago was the largest in recent times. The explosion that produced a tephra layer known as Mount Rainier C broadcast a wide arc of ash and small rock stretching north and southeast from the summit. In the Frozen Lake area, I always search for its popcorn-sized, sand-colored rocks riddled with tiny holes. They make a fascinating bookend with the surrounding landscape. These remains from an eruption just a moment ago in geologic time (2,200 years) sit scattered on and around Burroughs Mountain, a lava flow at the beginning of Mount Rainier's geologic story nearly a half million years prior.

Lava flows from this period built most of the mountain's summit cone that sits in the crater created by the Osceola Mudflow. About 1,000 feet high and a mile across, this recent explosive volcanism gives the mountain its current profile and height of 14,410 feet.

About a minute and a half ago on the 24-hour clock, or around 500 years ago, the mountain's most recent lahar sloshed down slope. The upper west flank near Sunset Amphitheater collapsed, activating a flow more than 100 feet high. Known as the Electron Mudflow, it rumbled 19 miles down the Puyallup River Valley and buried present-day Orting under a 16-foot mantle of debris.

In the day's final minute, eyewitnesses reported eruptions that included steam clouds, ashfall, and pumice. Credible reports of volcanic activity exist from the 1840s, but those between the 1850s and 1880s were probably steam clouds over the mountain or dust streamers resulting from upper slope rockfall. The *Seattle Post-Intelligencer* and the *Tacoma Daily Ledger* ran a series of stories covering Mount Rainier's eruptive activities during November and December of 1894, when observers reported small summit explosions. It's not altogether clear, though, whether the reports were geologically accurate.

Completing the 24-hour clock of 500,000 years of Mount Rainier's geologic history at 11:59:59 p.m., several matters need clarification. First, the frequency of events between 10:00 p.m. and the present time may lead readers to assume, almost certainly incorrectly, that more happened during contemporary times than in the distant geologic past. This apparent imbalance results from abundant evidence that supports recent events and the fact that proof of older ones tends to disappear because it gets buried or swept away by erosion, one of the mountain's most frequent and powerful landscape modifiers.

Readers familiar with the dramatic events at Mount St. Helens are likely to wonder about similar prospects at Mount Rainier. While the two volcanoes sit just 50 miles apart along the spine of the Cascade Range, their volcanic histories and eruptive behaviors differ markedly. Mount St. Helens tends to have slightly cooler and more viscous, silica-rich magmas that trap gases within them. This favors the growth of lava domes and

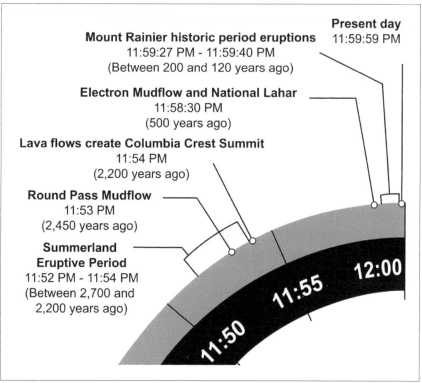

Just before midnight—a closer look at the sequence of some relatively recent geologic events at Mount Rainier. (See Appendix C for clock times and years of significant geologic events at Mount Rainier starting at midnight or 500,000 years ago.) *Kirsten Wahlquist*

eventually, large, explosive eruptions of pumice. Mount Rainier's magmas, on the other hand, flow hotter and more fluid, tending toward flows that travel longer distances without violent explosions. Only a few events at Mount Rainier compare with the spectacle of Mount St. Helens, with the Osceola Mudflow standing as a striking example.

Although it is unlikely that Mount Rainier will erupt in such eye-popping fashion as the 1980 Mount St. Helens eruption, geologists consider it among the most dangerous volcanoes in North America. Its numerous debris flows that include lahars and similar events over the last 10,000 years are its greatest hazard. Considering the 1.2 million people and property values exceeding $40 billion in close proximity to the mountain, it's no wonder that scientists, park managers, and down valley municipal officials fret over its destructive potential. Compounding their angst are other events already in motion in alpine environments worldwide; Mount Rainier's high country is no exception. The steady climb of global temperatures in the twentieth century has triggered vast changes to the mountain's glaciers and hydrology. Their long-term effects and the challenges they pose have already begun, setting researchers scrambling to understand and respond to them.

WARNING SIGNS OF CLIMATE CHANGE

Any account of Mount Rainier's geology is incomplete without a close look at the sprawling sheets of ice and snow that gird the mountain. Glaciers are the mountain's majestic white cloak, making it a prominent geographic beacon. They play fundamental roles in shaping and eroding the cone and as a major source for the Puget Sound area's groundwater.

Glacial and polar ice comprise Earth's largest freshwater reservoir and contribute to the health and maintenance of ecosystems that span from alpine heights to coastal lowlands. Wildlife, including the federally protected Chinook salmon and bull trout, pin their survival on the glacial meltwater that flows to the lowlands during the hot and dry summers. Pacific Northwest glaciers annually provide billions of gallons of water for generating electricity, irrigation, drinking, household use, recreational pursuits, and other purposes.

Glaciers are permanent bodies of ice that form over many years where the annual snowfall exceeds the melting rate. The process begins as light and porous newly fallen snow undergoes evaporation. The snowflakes gradually lose their geometric shape, become smaller and rounder, and pack closer together. This reduces the air space between them. Subsequent snowfalls add weight and pressure on the layer below, crystallizing the base layers to create a mass of solid ice. When it becomes so heavy that it yields to the pull of gravity, the mass has completed the transformation into a glacier. Essentially a river of ice, it flows over the ground on which it lays. Changes in the ground slope may cause the glacier's surface to crack. Crevasses and other features are the hallmarks of a glacier's relentless movement.

Glaciers are the original earth-moving equipment, having altered more than one-fourth of the planet's surface. Like ponderous landscape graders, they scrape weathered rock and soil. They grind resolutely at firm bedrock to transform Earth's crust to a fine sediment known as glacial flour. Heavyweight haulers, they carry everything from grains of sand to massive boulders many miles from their place of origin.

The changing size and movement of a glacier depends upon a variety of factors that include weather, climate, its degree of slope, and its exposure to sunlight. An increase in a glacier's mass is called accumulation. Loss of ice, snow, or water through evaporation or melting is ablation. The balance point between the two is its equilibrium line. It responds to climatic changes by moving up the mountain during warming trends and down during cooler periods.

The gain or loss of glacial ice and snow in a given year is a glacier's mass balance, usually expressed as water gained or lost. If we imagine that each glacier has a mass balance ledger book, years when accumulation exceeds melting register as positive years in the ledger. When it undergoes ablation, the ledger shows a negative year. A series of positive years correlates with a glacier's advance, and a negative series with its retreat. The well-documented rise in mean temperatures worldwide over the last hundred years has triggered a negative trend for most glaciers. Wholesale glacial retreat is underway, and the outlook is bleak. The Intergovernmental

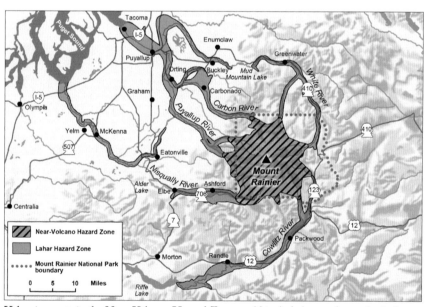

Volcanic events in the Near-Volcano Hazard Zone could include lava and pyroclastic flows, thick tephra, lahars, ballistic ejecta, and rock fall. Those in the Lahar Hazard Zone could include volcanic mudflows, potentially traveling long distances down the mountain's valleys. Not shown are the volcanic ash pathways that carry fine fragments of airborne volcanic rock downwind. *U.S. Geological Survey, redrawn by Kirsten Wahlquist*

Panel on Climate Change (IPCC), staffed by thousands of scientists from nearly 200 countries, provides international leadership on assessing climate change. Using the mid-range predictions for temperature increases, the IPCC estimates that the lower limit of perennial snow could rise an average of 1,500 feet on mountainsides around the world by the end of the century. In this scenario, all but the most immense glaciers would disappear. Seasonal water flow patterns, and the availability of water for human use, could change dramatically.

Glaciers in the Pacific Northwest get more attention from researchers than anywhere else in the contiguous United States. Under their watchful eye and careful measurement, scientists found that the 800 North Cascade glaciers—more than half of all those in the Lower 48—lost approximately 56 percent of their area between 1900 and 2009. Mount Adams, sitting 50 miles south of Mount Rainier, lost almost half of its ice in barely a hundred years. Like Adams, nearly all of the Cascade Range glaciers are retreating (losing area) and thinning (losing volume). The only exception is Mount St. Helens's Crater Glacier, the youngest in the Cascades. It began forming on the crater floor in the 1980s, where its steep sides shade the glacier nearly year-round as it collects snowfall and avalanches. Recession is the story on Washington's Olympic Peninsula, where glaciers lost between 17 and 24 percent of their volume between 1987 and 2010. The Olympic Mountains have lost one-third of their glacial ice area since 1980.

Mount Rainier's two dozen-plus named glaciers contain more ice than all the other Cascade volcanoes combined. They range from the diminutive Van Trump Glaciers that cover a scant 43 acres to the enormous Emmons that grinds and rolls over 4.2 square miles of mountain landscape. Joining it, the Winthrop, Carbon, Tahoma, and North Mowich Glaciers each have surface areas of nearly two square miles or greater. The mountain's glaciers total over 30 square miles, equivalent to half the area of Tacoma, and account for about 25 percent of the total ice in the conterminous United States.

Visitors are often surprised to learn that there is more to Mount Rainier's glaciers than their glistening white raiment. Debris from moraine deposits, rockfall, and wind-blown sediments accumulates on 19 of the mountain's glaciers, covering one-fourth of its total glacial area. Appearing as long, low piles of dirt and gravel, even a few inches can reduce melting by shielding them from direct sunlight. This insulates the glaciers, slows ablation, and ultimately, recession. The Carbon Glacier is the park's grand champion in total area of debris cover, catching so much rockfall off the precipitous Willis Wall that vegetation grows on top of it near its terminus. The Emmons, however, is the best example of how debris slows melting. In 1963, at least seven rock avalanches totaling 14.4 million cubic yards tumbled off Little Tahoma Peak to blanket the lower reaches of the Emmons Glacier. In subsequent years, the debris cover provided insulation that contributed to an advance and then later slowed a recession during a time of widespread glacial retreat.

Glacier studies are commonplace at Mount Rainier, the most glaciated peak south of Alaska. One project regularly records the volume of ice and snow of two glaciers on the mountain. Measurements began in 2003 and despite slight upticks in years of heavier snowfall, the data indicates large losses in mass balance. Since the study began, the Nisqually Glacier lost 102 million cubic yards of water equivalent, the measure of the amount of water contained within it. The Emmons lost nearly 195 million cubic yards during the same period. Park scientists estimate that between 1896 and 2015, the combined volume of all of Mount Rainier's glaciers diminished by 45 percent.

In a recent study, park geologist Scott Beason used three sets of aerial images to map the mountain's glaciers. After calculating the surface area of each glacier and perennial snowfield, he compared the results with those from five prior inventories stretching back over a hundred years. Beason found that glacial ice had undergone a 1.8 percent reduction between 2009 and 2015, the second fastest rate of decrease since the studies began. Overall, he found that the surface area of glacial and perennial ice at Mount Rainier shrank by 39 percent between 1896 and 2015. Beason believes that the pattern is significant and if the current climate trend continues, he expects more losses in the surface area of the mountain's ice. Small, lower elevation glaciers like the Stevens and the Van Trump could completely disappear within the next 30 or 40 years.

These and other glacial studies at Mount Rainier reveal a continuing trend, begun in the mid-nineteenth century, of the retreat and thinning of the mountain's glaciers. In other words, all of the glaciers' ledger books are in arrears, and all signs point to a continuing decline in area and volume. "Climate change is the elephant in the living room," said Paul Kennard, regional geomorphologist for the National Park Service, who studies Earth's physical features and their relationship to geological processes. Glaciers are excellent indicators of climate change because they respond quickly to changes in temperature and precipitation. The message deliv-

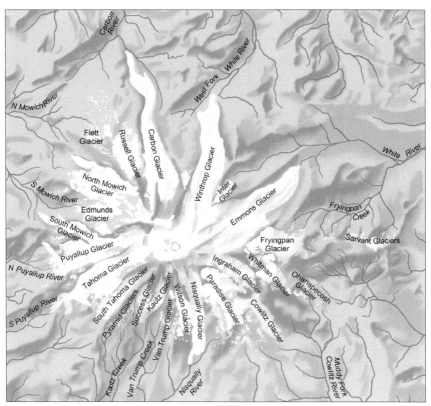

Aerial view of the mountain's major glaciers and rivers. *National Park Service, redrawn by Kirsten Wahlquist*

ered to researchers and park managers is loud and clear. "Based on the most up-to-date mapping, all of Mount Rainier's glaciers are at their historic minimums. All of the glacial terminus positions are at the highest recorded elevations since record keeping began," Kennard added. "Glaciers on the southeast and southwest sides of the mountain have suffered the greatest losses. We've lost eight glaciers in my lifetime."

Also lost are the fabled Paradise ice caves, a network of interconnected caves lying within the lower base of the Paradise Glacier. Spelunkers mapped two miles of passages and explored an additional mile of the ice caves that attracted hordes of curious visitors between the 1940s and the 1980s. The last of the caves disappeared in the early 1990s, marking the end of an extraordinary way to experience the mountain's glacial system.

From Mountain Heights to Salish Seas

Continuing their 150-year retreat, Mount Rainier's glaciers present complex problems that scientists, resource managers, and public officials are just beginning to understand. Changes to both alpine and downslope ecosystems that are already underway will likely be far-reaching and long lasting. Plants and animals that depend on the annual cycle of glacial meltwater that buffers summer droughts will need to adapt to less water during those times and to more rain in winter. The reduced summer flow will challenge municipalities to provide drinking water and hydroelectric power.

In addition to altering annual flow patterns, retreating glaciers create other circumstances that concern geologists and park staff. Most glacial movement happens during the warmer months as ice flows downhill, on average a couple of feet per day. But when observant climbing rangers noticed that stakes used to measure mass balance on the lower Nisqually Glacier did not appear to be moving very much, the park's geologists sprang into action. As glaciers retreat, they sometimes abandon ice masses in the lower valleys. Scientists call them stagnant ice because they no longer move at the same rate as the retreating mass. This sediment-rich ice stores huge volumes of water beneath it that can release suddenly and without warning as a glacial outburst flood. Also called a *jökulhlaup* (yo´-kull-lup), Icelandic for "glacier burst," they can cause considerable damage as they move down valley. Unrelated to volcanic activity, these events give no advance notice on seismographic instruments. They move rapidly upon their release, posing extreme danger to everything in their path. These powerful surges of water scour the river's channel walls and bed, incorporating vast amounts of sediment. This speedy discharge, set in motion by water, then becomes what geologists call a debris flow.

Knowing that the road leading to Paradise sits about a mile below the Nisqually's barely-moving ice and that Longmire is another four miles downriver, geologists feared that if the lower glacier was becoming stagnant that a glacial outburst flood could jeopardize lives and property. Knowing the rate of movement of the glacier's surface can indicate whether stagnant ice may be present and if outburst floods are possible. Measuring a rock's position relative to a fixed point over time gives the distance that the rock moved because of glacial flow.

After two seasons of measurements, the data suggested that the upper glacier moved at a normal velocity of about two feet per day. The middle section appeared to be in a transition zone with some movement, but the entire lower area crept just two inches or less per day. The ice was stagnant, the conditions right for an outburst flood. The problem was, no one could predict or prevent it.

No fewer than nine outburst floods have raced down the Nisqually River in recorded history, the most recent one in 2012. A 1955 outburst caused flooding at Longmire. During the twentieth century, more than three dozen outburst floods originating from five different glaciers have damaged roads, bridges, trails, and visitor facilities. The majority gush from the Nisqually, Kautz, and South Tahoma Glaciers. Located on the mountain's south side, their increased exposure to solar radiation and to storms, plus their slope angle, makes for more of these floods than anywhere else at Mount Rainier.

The park's best-studied glacial outburst floods originate at the South Tahoma Glacier, where water has rushed down valley 30 times since the 1960s. Capable of releasing more water than a hundred-year flood, they typically take place during spells of hot, dry weather or heavy rainfall. And they carry more than water, too. In one seven-year span, over 30 million cubic yards of sediment moved down Tahoma Creek, nearly enough to fill two lines of bumper-to-bumper dump trucks stretching from the glacier to New York City. The flows have been so powerful that they've created three different channels in recent years. One flow moved the creek 500 feet closer to the Westside Road, increasing the chances of flooding and erosion. Flows have had considerable impact on the old growth forest bordering the creek. Standing in a dry, six-foot deep channel, I marveled at the remaining trees, their trunks battered by uncountable boulders the size of bear cubs. These occasional inundations have left more than 500 trees standing dead here, killed either by root suffocation due to sediment loads, drowning from excess water, and in some cases, from being cut off from their water supply. These towering, silvery snags create a ghost forest, a grim reminder of the extended effects of debris flows. Other local damage included the destruction of a picnic area and campground, a section of trail, and repeated washouts of the Westside Road. Flooding became so common that following an environmental assessment in 1992, the park closed the road to vehicular traffic beyond Dry Creek.

People have escaped injury in outburst floods thus far, but some have come away with hair-raising tales. In one incident, a ranger reported the ground shaking violently, and a thunderous roar. In 2015 a visitor shot a cellphone video as a flow thundered within a few feet of where he stood. Nearby, a park volunteer sought refuge on higher ground as boulders churned and tree trunks snapped around her.

I once tagged along with park geologists on a day trip to upper Tahoma Creek. One of the crew was developing a new procedure using geochemical methods to measure stagnant ice. Taking water samples over extended periods that measure pH, temperature, and other parameters, researchers hope to identify changes in the movement of glacial ice. We traipsed up and down unstable moraines during one of the hottest days of the year in the midst of an August heat wave—prime time for outburst floods and debris flows. Although park policy forewarns hikers to head for high ground in the event of an outburst flood—often signaled by rumbling sounds like those of an approaching train—geologists have an ironically different notion. They envy colleagues who have witnessed glacial outburst floods, and have a standing joke that, given the chance, they would actually run toward it.

Tahoma Creek, the site of numerous glacial outburst floods in the last 50 years. Notice the dead and battered standing trees in the background and the braided channels in the foreground. *Scott Beason/National Park Service*

No one witnessed the park's largest debris flow in recorded history, but imagine the reaction when employees encountered the results on their way to work in early October 1947. East of Tahoma Creek on the Nisqually-Paradise Road, a monstrous flow ran wild after two days of heavy rains, causing the lower mile of the Kautz Glacier to collapse, and triggering a glacial outburst flood. It quickly became a debris flow and in a series of pulses blanketed the area with nearly 50 million cubic yards of material to a depth of 28 feet—about the height of a utility pole. The remaining snags, their roots suffocated by sediment, mark the spot near the Kautz Creek Picnic Area.

Evidence of more recent flows on the mountain's south side lies just upstream of the Cougar Rock Campground and Picnic Area where Van Trump Creek runs near the road. Beginning in 2001, a series of four debris flows in five years deposited over 490,000 cubic yards of material into the Nisqually River. The sediment from the 2005 flow alone contributed to a five-foot rise in the riverbed. More dead trees and a river lying higher than the adjacent road bear testimony to a 38-foot rise in the riverbed since the road was built over a hundred years ago.

Van Trump Creek underscores the increasing frequency of flows within the park, and scientists anticipate more as one effect of climate change. As glaciers retreat mountain wide, their ice no longer serves as the "glue" that holds and consolidates drift (glacial sediment and rocks), making it available for downslope transport. The drift becomes susceptible to slope failure, landslides, and when moved by water, becomes a debris flow. Knowing that flows can begin spontaneously, park staff felt the need to prepare.

The program's pilot site was a washed out, 700-foot section of the Westside Road alongside Tahoma Creek, about a mile north of the Dry Creek gate. After site evaluations and planning, structures made of large woody debris were installed along the road edge into the creek and deeply anchored with massive rocks. Imitating natural logjams, the structures dissipate a stream's energy by slowing its flow, collecting sediment, and protecting the roadbed. The final step used bioengineering techniques to construct a low, woven wall of Sitka willow and red alder cuttings to help hold soils and rebuild the creek's riparian zone. Park managers are optimistic that this fresh approach will protect the road and bode well for dealing with other geologic threats.

Terms used to describe debris flows and others like them have changed over time. "Mudflow" was popular among earlier researchers, but the technical differences between them and debris flows are so slight that most experts now refer to all such events as debris flows. Despite the shift in nomenclature, the Kautz, Electron, and Osceola Mudflows retain their original names. Some media outlets also prefer to use "mudflow," a term the public easily understands. Today's geologists know the Osceola, Electron, and events like them as lahars, defining them broadly as any debris flow that begins on the side of a volcano. By this definition, all Mount Rainier debris flows qualify as lahars. Park managers, however, use "debris flow" for those that are small enough to remain within the park and use "lahar" for those that leave park boundaries.

A lahar is a slurry of water and sediment that roars down a mountainside "like a thousand stampeding water buffalo"—the word's translation from Indonesian into English. Over 60 such stampedes have charged down Mount Rainier's slopes in the last 10,000 years. Lahars can travel up to 60 miles per hour on steep slopes and could overwhelm parts of Tacoma or the Seattle suburbs in little more than an hour. They conform to the unrelenting laws of physics, flowing downhill and within their river valleys until making the lowlands. There, they spread widely. The only way to escape is to move to higher ground—immediately. Mount Rainier averages one lahar every 300 to 1,000 years, the most recent being the Electron Mudflow a little over 500 years ago. Although eruptive activity causes lahars more frequently, they also can occur during volcanically quiet times, heightening their unpredictability.

Geologists classify lahars as either non-cohesive or cohesive, a distinction especially important to city officials and those working in the emergency response sector. Non-cohesive lahars begin as watery surges and accumulate sediment during their downstream passage, but change gradually to a dilute flow. Cohesive lahars form on steep mountainsides where the hot and acidic waters of the mountain's hydrothermal system chemically transform the rock to clay and other minerals. The changes allow the rock to absorb and hold huge quantities of water, which weaken it. Eventually, the rock mass collapses and gives way to gravity. The slippery, clay-rich, cohesive lahars tend to have more volume and cover larger areas than non-cohesive ones, presenting additional challenges for disaster response teams. The Osceola Mudflow, a cohesive lahar that tore down the mountain 5,600 years ago, was estimated at five times more mobile than a volcanic avalanche of similar size.

Whether as glacial outburst floods, debris flows, or lahars, geologists believe that these types of events will continue as the primary, regularly occurring, but hard-to-predict geologic hazards at Mount Rainier. In spite of the technical differences between the phenomena, their common denominator lies in the water and sediment they carry away from the mountain. Sediments suspended in water range in size from silt and sand to gravel, cobbles, and boulders, and it all has to go somewhere. For each of the mountain's glacially spawned rivers—except the Cowlitz, which empties into the Columbia River—that somewhere is Puget Sound.

While dramatic, sudden debris flows unpredictably punctuate the mountain's geologic history, the steady, inexorable movement of sediment happens every day. To imagine Tahoma's rivers as turbulent, serpentine conveyor belts that haul a million tons of sediment to Puget Sound each year is to begin to understand aggradation, one of the growing issues facing the region. In contrast to the largely unpredictable dangers of cataclysmic flows, aggradation progressed so gradually during most of the twentieth century that it gave scant notice of the wholesale changes underway on the mountain and in the lowlands. The mountain's rivers showed little increase in their sediment loads between the 1960s and the 1980s, but things changed abruptly beginning in the 1990s. Debris flows and seasonal floods boosted sediment loads and widened rivers between 1994 and 2009.

Aggradation is a long-term hydrologic process that under normal conditions supplies sediments and vital nutrients to beaches, river deltas, and other coastal areas. It helps build and sustain a range of ecosystems that support a broad diversity of plant and animal life. Excessive amounts of sediments, though, make it difficult for fish to breathe and navigate. Sediment-laden water chokes off the oxygen supply to tree and shrub roots when it overtops the river's edge.

Most of Mount Rainier's rivers are "braided," consisting of frequently changing channels separated by gravel bars, and characterized by steep grades and high sediment loads. A key piece of the aggradation puzzle is that braided rivers typically receive more sediment than they can remove. Sediment settles onto and builds up the riverbed, effectively "raising the floor" of the river, leaving less space for water to run through the channel. During high flow events, water spills from the river's bank to cause flooding or widens the channel through erosive action.

No one knew much about aggradation until the 1980s, and even less about the rates at Mount Rainier. That changed, when as a graduate student, now-park geologist Scott Beason undertook a comprehensive study of the mountain's rivers. Leading a team of scientists from cooperating agencies, Beason used electronic surveying instruments, GPS technology, and historic maps to calculate aggradation rates between 1910 and 2000. The results startled everyone. Sediment deposits had averaged four inches per decade in Mount Rainier's rivers during most of the twentieth century. But the rate skyrocketed to three feet between 1996 and 2006, an increase by a factor of nine.

Beason and his colleagues conducted follow-up surveys to keep an eye on aggradation in the park's rivers by focusing on four locations on the Nisqually River and another on the White. Their results showed continuing accumulations at some of the sites. One of the highest amounts was at Sunshine Point, about a half-mile from the Nisqually Entrance. At the site where a campground and picnic area were swept downriver in the 2006 flood, 35,000 cubic yards of sediment were added between 2005 and 2012. Making matters worse, the Nisqually riverbed tilts toward the road there, putting it at risk to subsequent washouts.

More problems loom further upriver in the Longmire area. Although the crew found only a slight increase in aggradation over a 15-year period there, the river sits 29 feet above Longmire Village, endangering park infrastructure. Flooding events that overtop the Nisqually's banks can rush downslope to damage roads, buildings, vehicles, equipment, and historic buildings. Despite finding only a gradual uptick in the aggradation rates near Longmire, Beason expects an increase in coming years.

On the mountain's north side, aggradation dumped more than 70,000 cubic yards of sediment into the White River in seven years, double the input at Sunshine Point and equivalent to two concrete trucks dumping their payloads into the river every day during that time. The hotspot lies on State Route 410 near mile marker 59, where the influx of sediment—which at one point accumulated at nearly five feet per decade—has raised the riverbed 12 to 16 feet higher than the roadway. This makes the road vulnerable to washouts where flooding has already damaged some of the adjoining old growth forest. Several inches of sediment cover the forest floor, affecting its ability to slow and absorb water from bank overtopping. The sediment also chokes the expansive root systems that run a foot beneath the surface, just as it did in the Tahoma Creek and Kautz Creek flows.

As a river flows away from a mountain, its gradient decreases. It drops coarser materials like gravel and cobbles onto its riverbed, but fine particles like silt and sand continue downstream. This creates problems that extend far beyond the river. Rivers draining Washington's Cascade Range move over six million tons of sediment annually into Puget Sound; Mount Rainier's rivers account for a million. Sediment entering Puget Sound means more flooding, which has increased in the Carbon, Nisqually, Puyallup, and White River basins. Since 1990, ten floods resulting from sediment-clogged rivers have become presidentially-declared national disasters.

The record increase in sedimentation and flooding presents substantial challenges for public officials in neighboring King and Pierce counties. The once-common practice of dredging strategic portions of river to remove sediment is no longer practical. Federal law protects critical habitat for steelhead, Chinook, and coho salmon. The difficulty in obtaining permits and the temporary solution of flood risk reduction through dredging has forced officials to look elsewhere for answers. The King County Flood Hazard Management Plan, for example, projected expenditures of over $398 million between 2012 and 2018 for flood control and floodplain management in an effort to reduce potential damage. Similar efforts in the Puyallup River watershed, which lies mostly in Pierce County, show promise of reducing flood risks.

Even with these preventive measures, people living near Mount Rainier's rivers and the Salish Sea margins will continue to experience flooding in the coming years. Near the Seattle suburb of Auburn, scientists predict a 50 percent increase in sediment and other rock debris in the White River over the next fifty years. These coarse particles moving along the river bottom will add to the accumulation already underway and increase flood risk in the area.

Moreover, a strong correlation between air temperature and the amount of suspended sediments in rivers indicates a likely relationship between increasing temperatures (resulting from climate change) and aggradation. As the climate continues warming, geologists warn that aggradation rates may similarly increase.

Given Mount Rainier's geologic history as a dynamic, ever-changing landscape, one might expect that people throughout the ages would have kept a safe distance from it. Native American oral tradition, along with a growing body of archaeological evidence, demonstrates that people have ventured onto the mountain for thousands of years. Building upon the geologic foundations and processes that outline the place, we now turn our attention to human activity at Mount Rainier.

2

The People: Footprints of Days Past

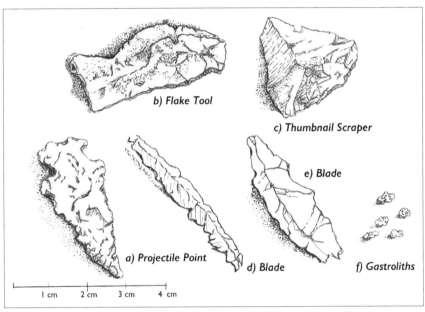

b) Flake Tool

c) Thumbnail Scraper

e) Blade

a) Projectile Point

d) Blade

f) Gastroliths

1 cm 2 cm 3 cm 4 cm

A selection of flaked-stone tools from a subalpine site at Mount Rainier. These are among the thousands of artifacts and tool-stone fragments found in the park. Those pictured range between 2,000 and 5,000 years old: a) projectile point; b) flake tool; c) thumbnail scraper; d) microblade; e) macroblade; and f) gastroliths, the small, polished stones found inside the gizzards of grouse and their relatives. *Drawing by Lucia Harrison*

People on the Landscape

The vine maples demand my attention this September morning, ablaze in a fiery display of fall glory. Scattered on the green hillsides, the oranges and rusts, ochres and crimsons signal the turning of the season as I arrive at the park's most recent archaeological excavation.

Prior to the installation of new underground utility lines in the Ohanapecosh Campground in the park's southeast corner, cultural resources staff tested for archaeological artifacts. The first people who came to Mount Rainier hunted and gathered with stone tools that they used to cut, pierce, and scrape game and other materials. Sharp-edged stone tools work well, but are brittle and fragile, requiring frequent repair and replacement. Wherever people made, used, or repaired these tools, they left behind a shower of chipped, fine-grained stone fragments. These remains became clues for archaeologists about the people that left them.

Park staff recovered stone tool remnants during initial testing at Ohanapecosh, and when sampling the landscape more broadly, they found more chipped tool remains at over a dozen locations, probably small campsites. Dating them was easy because of Mount Rainier's unique layer cake stratigraphy. Numerous eruptions at the mountain—and from other volcanoes—have deposited ash and other volcanic material in layers on the ground. These layers relate to the known age of volcanic events and thus the approximate age of the tool-stone fragments contained within them. The earliest Ohanapecosh sites sit below the Mount Mazama ash layer, left by the eruption that created Oregon's Crater Lake about 7,700 years ago, indicating that people occupied the sites at about that time.

Although the Ohanapecosh finds are some of the oldest at Mount Rainier, they matter for another reason. Most archaeological sites at the park occur at 4,000 feet or higher in elevation, but those at Ohanapecosh are among the first precontact sites found below 2,000 feet. Precontact refers to the period prior to Native American encounters with European and Russian explorers in the late eighteenth century.

Park archaeologist Greg Burtchard greeted me when I arrived at the Ohanapecosh Campground. This would be my third time assisting on excavations, and Burtchard had spent considerable time helping me understand the park's archaeological record—all of the physical evidence of peoples' presence at Mount Rainier. This record includes camping and butchering sites, fire hearths, stone tool artifacts, and other clues of human presence. After introductions and a quick tour of the project, Burtchard put me to work. From one of the original sample holes, the crew had excavated a unit that measured over nine feet square and over three feet deep.

Ohanapecosh Soil Stratigraphy. In an archaeological unit in the Ohanapecosh Campground, the tephra layer sequence extends back nearly 8,000 years: 1–Mount St. Helens eruption, 1980; 2–Mount St. Helens "X" eruption, 1837; 3–Mount St. Helens "W" eruption, 1472; 4–Mount Rainier "C" eruption, about 2,330 years ago; 5–Mount St. Helens "Y" eruption, about 4,270 years ago; 6–Mount Rainier "L" eruption, about 7,390 years ago; 7–Mount Mazama "O" eruption, about 7,670 years ago; 8–over-bank flood deposits, between 8,070 and 10,570 years ago. *Greg Burtchard, Mount Rainier National Park archaeologist (retired), modified by Kirsten Wahlquist*

Standing chest deep in the unit, one of the researchers carefully skimmed and scraped away small bits of dirt, hoping to locate artifacts in situ. She brushed the loose material into buckets and handed them up and out of the unit for screening. I helped the others shake and sift the dirt through fine wire-mesh screens, looking for any pieces of chip stone tools.

The day's finds were few and unsensational in appearance, little more than fingernail-sized fragments of tool-stone material. Over lunch though, Greg showed me two intriguing items. One was a broken projectile point of a style typically associated with cultural deposits ranging between 7,000 and 9,000 years old. Technically not arrowheads, since bow and arrow technology did not gain widespread use until around 3,000 years ago, projectile points like this were affixed to small shafts to make darts. Socketed into longer shafts, the darts were launched with atlatls. Atlatls served as a fast-moving extension of the throwing arm, delivering about 60 percent more thrust than the conventional means of throwing a spear, much like the plastic ball throwers that people use today to play fetch with their dogs. The other was a stone scraper about the size and shape of a large guitar pick. Perfectly carved and unblemished, the milky white stone looked as if it had never scraped bark off a twig or fat from the inside of an animal's hide. Holding it, I tried to imagine its origins 7,300 years before. How did the toolmaker acquire the source stone? Where did the artifact take shape— around a nearby campfire? How was it dropped, lost, and forgotten?

The crew had also discovered two ancient fire hearths. Fires serve and unify groups as they provide heat for cooking, radiate warmth, and shed light for tool making and other activities. Food scraps like seeds or bones in the charcoal remains tell stories about the people who gathered there. The charcoal itself, through radiocarbon dating, pinpoints the age of the hearths.

For Burtchard and his crew, the challenge—and the reward—of their work comes from working out the complex puzzles of how people used these areas over the ages, and from seeking the highest possible degree of scientific validity so that their conclusions transcend mere conjecture. Excavations at the Ohanapecosh Campground sites suggest that Native Americans, possibly of Yakama, Klickitat, or Upper Cowlitz ancestry, made short-term camps here beginning at least 8,000 years ago. They were passing through, travelers on their way to somewhere else, probably to Mount Rainier's resource-rich mid-elevations or back to their lowland villages.

SALISH SEA NEIGHBORS

Two barriers stand in the way of a full understanding of the people living near Mount Rainier prior to settlement by European Americans in the nineteenth century. First, following contact with outsiders that began on Vancouver Island in the 1770s, a series of epidemic diseases sickened and killed native people throughout the Western Hemisphere, eventually reaching Pacific Northwest Indian communities. Estimates of the population loss from introduced diseases run at a minimum of 80 percent mortality, higher than Europe's Black Death plague in the fourteenth century. Diseases caused an unrivaled degree of suffering and cultural disruption among the area's Indian people. Smallpox and malaria were the deadliest, although measles, tuberculosis, and influenza also decimated populations. Among the Puget Sound region's Native American groups, smallpox epidemics in the 1700s and again in 1853 were the bookends to measles and influenza outbreaks that altogether caused a 63 percent population loss in less than a century. These catastrophic events compromised the attempts of anthropologists and historians to chronicle Indian ways from the outset, for the culture had already been permanently altered.

Second, archaeological finds at Mount Rainier and across the Pacific Northwest are difficult to come by. Most of the rivers flowing from Mount Rainier have active floodplains. Frequent floods and geologic events like volcanic eruptions and debris flows disturb or completely cover archaeological evidence. Heavy undergrowth in the temperate maritime forests covers most signs of human presence. In addition, the forest's acidic, moist soils quickly decompose artifacts made of wood, bark, or other organic materials that could leave clues about human activity.

Despite these limitations, tribal stories, detailed interviews of a few tribal elders, and a growing archaeological record permit at least a glimpse of precontact Native American life around Tahoma.

More than 50 named groups of Native Americans lived around Puget Sound in the 1850s. The original inhabitants spoke one of the Coast Salish languages, part of the Salishan language family. People living in the Hood Canal area and its drainages spoke Twana. Those along Puget Sound and in the drainages flowing into it spoke Puget Salish, also known as Lushootseed—a word derived from one meaning "salt water" and another meaning "language." Comprised of many dialects, Lushootseed was the region's principal means of communication. The Muckleshoot people living in the White and Green River Valleys northwest of Tahoma call their southern dialect Wuhlshootseed. My Wuhlshootseed language teacher told me that her ancestors would have been able to talk with other tribal people, including the Duwamish, Puyallup, Nisqually, Snoqualmie, and Squaxin. They could also speak with the northern dialect people, including the Tulalip and others. Some people probably spoke both Twana and Lushootseed. Those living along the Strait of Juan de Fuca, around Bellingham Bay, and further north spoke dialects of the Straits Salish language. Still others spoke Sahaptin, the principal language of people living east of the Cascade Mountains.

Throughout most of the nineteenth century, native people and European Americans communicated with each other through widespread use of a pidgin language, the Chinook Jargon. Its predecessor originated in the late 1700s when fur traders and local people on the west coast of Vancouver Island needed a way to barter with each other. The jargon they developed gradually made its way down the Pacific Coast, evolving into

Chinook Jargon and first appearing in a written form in Lewis and Clark's journals at the mouth of the Columbia River in 1805. The Chinook Jargon retained some of its northern origins but borrowed heavily from the Lower Chinook language of the Lower Columbia River. Words from French, English, and other languages expanded the jargon to around 250 words by the early 1880s, its period of peak usage. Of the many geographic features at Mount Rainier that derive their place names from Chinook Jargon, a few include Ipsut (hidden) Pass, Mowich (deer) Lake, and Tipsoo (grass) Lake. In addition to shared languages, other cultural elements linked the area's Native Americans. People enjoyed a varied diet of game, plant, and marine resources, but the five types of salmon that seasonally crowded the river sustained their lives and strengthened their culture for the last 3,500 years or more. Fresh salmon was widely roasted and eaten, and the ability to preserve it through smoking and drying carried families and entire villages through the lean winter months.

Known to some native people as *Three Jumps* or *Lightning Following One Another*, salmon occupied—and continues to occupy—a sacred space in their culture. It resides in the people and the people in it; they are inseparable. Native cosmology centered on a belief that all parts of the Earth possessed conscious spirits. Each plant, animal, and mineral held a spot as a valued member of Earth's community. As a sentient being, the salmon offered itself as a gift to the people who in return performed ceremonies to express their gratitude so that it would return in abundance the next year. People offered prayers and placed the fish's bones in the river so that it would revive itself, swim home, and come back next spring. First Salmon ceremonies continue to be held today among many Northwest Coast tribes.

The custom of intermarriage also connected Coast Salish people to each other. Most groups encouraged people to marry outside their own village, which effectively expanded mutual support networks to help people weather stressful times. Such networks assured safe passage and encouraged peaceful relations by enlarging one's circle of relatives. Marriages across cultures were also commonplace and yielded similar results.

Finally, their connection to the land bound people together. Watersheds—defined as specific areas bounded by ridges and draining into major rivers—did not provide formal political boundaries, but served as organizing structures of social unity. The people of the White, the Cowlitz, the Nisqually, the Puyallup, and other rivers were joined together by the watershed in which they lived. They depended upon its waters for transportation and subsistence. Sovereignty for the Coast Salish people grew not from maps, constitutions, or laws, but from a living relationship with the land and the river as they fished, hunted, and gathered to fashion their lives.

Within their watersheds, people lived in villages, the main organizational unit of indigenous people for the last 3,000 to 4,000 years. Rather than identify as members of large groups we might call tribes, most native people thought of themselves as belonging to independent villages. These sites typically were located along waterways that served as transportation routes, often with beach access, with fresh water nearby for household use. The sites provided sufficient economic reasons for people to live there, usually ready access to fish, game, and plants. In areas susceptible to occasional raids by the more warlike people from the north, escape routes into the forest provided for peoples' safety.

Villages consisted of one or more large, permanent plank houses or longhouses, and often one or more smaller structures. Longhouses usually faced the water in a single row, each up to 200 feet long by 50 feet wide. A longhouse often endured for generations and symbolized a family's lineage and history, serving as a physical and spiritual anchor. A man and his wife or wives and their children shared a longhouse with other families, usually relatives. Kin that spanned multiple generations, extended family, and slaves (if the family was wealthy) all lived under the same roof. People partitioned their living spaces by hanging woven mats from beams or using planks or stacked boxes as dividers. Each living space usually had its own fireplace. Although there was no formal village council or official headman, the head of the wealthiest household was the village leader.

Arborvitae, the Tree of Life

Another similarity among the region's indigenous people was their use of, and dependence upon, western red cedar—*arborvitae*, the tree of life. In every spiral of life, from an infant's first cry to an elder's last breath, cedar wove itself into the lives of the first people. Of cedar, renowned Haida carver and artist Bill Reid wrote, "Oh, the cedar tree! If mankind in his infancy had prayed for the perfect substance for all material and aesthetic needs, an indulgent god could have provided nothing better."

Infants were born onto mats of woven cedar bark strips and then wrapped in the soft shredded inner bark. In some villages, family members placed the afterbirth in a large cedar stump to ensure a long life or in a cedar's crown to endow the child with bravery. Blankets made from the inner bark,

Woven in many sizes, styles, and shapes, cedar baskets were highly versatile and used by Coast Salish people in a wide variety of ways. Pictured here: a) split cedar branch; b) bear grass; c) unprocessed cedar root; d) diagonal warp twining clam basket made by Celia Jackson, Suquamish; e) coiled cedar-root berry basket with bear grass by Lucy Riddle, Suquamish; and f) coiled cedar-root cooking basket, owned by Jacob Wa-haleh. *Drawing by Lucia Harrison*

beaten with rocks and hand-worked until soft, swaddled the infant. People made diapers from the same material. They made cradleboards of cedar, too.

On cold mornings, a cedar hearth board and drill used friction to kindle cooking fires. Women added hot rocks to a tightly woven cedar basket to boil water for soup or stew. People used cedar dishes or bowls to serve salmon grilled on cedar barbecue sticks.

As children, girls learned to gather and prepare cedar bark and roots in the springtime. Their elders taught them to weave strips of inner bark and roots into mats, baskets, hats, or other clothing. In summer, they dropped freshly picked huckleberries into baskets before drying them on cedar racks. Boys learned to hunt with cedar spears and arrows.

On the water, several types of river and oceangoing canoes provided transportation over both short and long distances. Expertly crafted from cedar logs, they were swift and carried heavy loads. Men built fish traps—weirs—of cedar lattices lashed together with withes. The weirs spanned the waterway, allowing fish to enter but not to escape. Standing with dip nets on dangerous, narrow platforms, men harvested the fish. Cedar fishing floats fixed onto nets, cedar fish clubs, and cedar herring rakes helped with the various harvests.

People fashioned cedar into string, twine, or rope. Without metal nails, it bound mauls to hafts, rungs to ladders, and found many other everyday uses. At some celebrations, dancers wore elaborate cedar masks while cedar rattles or drum logs produced the necessary rhythm.

Cedar cured ailments, too. Chewing the buds relieved toothaches and sore lungs. People soaked their sore feet in hot footbaths that contained yellow cedar branch tips. Sweat baths and steam baths that used cedar branches laid upon hot rock or coals, with water poured upon them, helped cure rheumatism and other illnesses.

At the end of life, in some places people swept the house's walls with singed cedar branches as a form of purification. Others burned branches to frighten away the ghost of the deceased. They laid their dead to rest in cedar coffins or dugout canoes containing prized possessions, then hoisted it onto a mortuary pole or stilts or lashed it to a tree. The remains bore witness to lives intertwined with Long Life Maker, western red cedar.

Known variously as canoe cedar, shinglewood, and arborvitae, western red cedar grows mostly along the Pacific Coast. It ranges from southern Alaska and British Columbia through Washington and Oregon to northern California and east into northern Idaho and western Montana. Along the coast, it grows from sea level to 4,000 feet in elevation. Under ideal growing conditions, it stretches up to 200 feet skyward and 16 feet in diameter. Some of the largest trees at Mount Rainier are in the Grove of the Patriarchs, a one-mile hike along the Ohanapecosh River on the park's east side. Western red cedar's conical form, fluted base, and gracefully drooping branches usually identify it at a distance. Up close, inspection of its vertical, fibrous bark and scaly leaves removes any doubt about the tree's identity. Unlike other conifers, its leaves display in flat, braid-like sprays.

Similar to western red cedar is yellow cedar. Also a member of the cypress family, it is called Alaska yellow cedar, Sitka cedar, and yellow or Alaska cypress. One foolproof way to distinguish red from yellow is to stroke the fronds backward from the branch tip toward the trunk. Yellow cedar is prickly; red is not. Even though it doesn't lend itself to raising longhouses or carving canoes, yellow cedar was—and remains—highly valued by native people. It is prized for making bows, masks, canoe paddles, and many other everyday items. Yellow cedars live longer than any other trees in the region. A small stand about a mile below Ipsut Pass in the park's northwest corner is about 1,200 years old.

Both red and yellow cedars have a tight, straight grain that makes for trouble-free splitting and a soft texture for easy carving. The outstanding property distinguishing red from yellow lies in its natural preservatives. A group of chemical compounds called the thujaplicins comprise between 10 to 20 percent of red cedar heartwood, giving the wood its natural durability. These chemicals inhibit the growth of many bacteria and are toxic to fungi that decompose wood. Essentially natural oils, they prevent rot and decay, making it the region's most reliable and longest lasting wood for building and carving.

Cedar continues to play a central role in the lives of native people in the Pacific Northwest. On several occasions in late winter or early spring before the sap began running, I accompanied people into the forest to collect bark for weaving. One day, I helped a master weaver harvest cedar bark from trees scheduled for clearing for a building project. First, we found the side of the tree with the fewest knots or branches. We used a hatchet to make a horizontal incision six to 10 inches wide in the bark low on the tree. Then we pried and pulled the strip of bark up the sides of the tree, 20 feet or more. We pulled dozens of bark strips before setting about the difficult task of separating the inner from the outer bark. Finally, we rolled the strips of inner bark into bundles and tied them off with scraps. That evening we delivered the bundles to a tribal elder, herself a master weaver. Her eyes twinkled as we unloaded our cargo and set it to cure before further processing and weaving. "We are so rich," she exclaimed. "Look at all this. We are so rich!"

Nowhere is cedar's legacy more evident than in the tribal canoe journeys that have enjoyed a resurgence in recent years. Each summer, a different Salish Sea tribe hosts a gathering that convenes thousands of people for five days of ceremony and celebration. At the recent Paddle to Lummi near Bellingham, Washington, over 100 oceangoing cedar canoes, built as masterfully as those of long ago, made landfall at the Stommish Grounds. Paddlers wore splendid cedar hats and vests as they performed traditional songs and dances, exchanged gifts, and participated in other cultural activities.

THE MYTH:
"INDIANS WERE VERY SUPERSTITIOUS AND AFRAID OF IT"

When European Americans began settling the Puget Sound region in the 1800s, they encountered Native American villages at strategic locations along the major waterways. Long-standing Indian settlements were numerous at present-day Seattle, Tacoma, Olympia, and hundreds of other locations. There was little knowledge, however, of Indian travel into the mountains.

Native stories tell of people in the mountains in general and at Tahoma specifically, but accounts differ regarding the heights to which they ascended. Nisqually tribal member, historian, and writer Cecelia Svinth Carpenter told of a boundary "where trees stop growing and eternal snowfields lie deep...the sacred demarcation line that encircles the entire mountain." Nisqually people did not pass beyond this line. It was a combination, Carpenter wrote, of respect for and fear of *Ta-co-bet*, the Nisqually name for the great white mountain, which kept people from traversing its uppermost flanks.

Other stories present differing views of the Indian relationship to Tahoma. Arthur C. Ballard grew up in Auburn, Washington, not far from Muckleshoot villages on the Green and White Rivers that had stood for generations. His lifelong passion for language compelled him to translate stories he heard from Indian elders. He recorded and translated two versions of "Young Man's Ascent of Mount Rainier." The first version begins with the words, "This story is not a myth. The man in this story was a real man." The second version begins, "The grandfather of my grandmother went up to the summit of Mount Rainier." In the stories, the young man finds a lake at the summit while searching for magic powers. A lake does, in fact, exist in an underground cave on the mountaintop.

In 1886, a young European American named Alison Brown accompanied a group of about 30 Yakama Indians on a hunting expedition up Cowlitz Divide on the mountain's southeast flank. Failing to find any game, they continued upward until seven or eight of the group—with Brown in tow—decided to climb toward the summit. Brown later said, "We did not try to reach the highest pinnacle," but on their descent spent the night at the base of Gibraltar Rock.

These accounts, supplemented by extensive archaeological remains and the well-known Indian trails system, clearly show that for ages people were drawn to the place "where the waters begin." Ignorant of such evidence, the newcomers assumed that Native Americans avoided mountainous areas, and believed that primitive superstitions restrained Indian people from venturing into the mountains.

Historians and anthropologists now believe that travel by native groups was sufficient to create and maintain routes that linked lowland areas to the high country. There is evidence of frequent and long-lived travel between the eastern and western sides of the Cascades and that some of today's trails, roads, and highways follow these earliest pathways. In addition, Native Americans contributed substantially to the successes achieved by early European American explorations of Mount Rainier. A rich historical record details Indian involvement in a variety of adventures.

Take the example of Dr. William Fraser Tolmie. Freshly graduated at age 20 as a medical doctor and surgeon from Scotland's Glasgow University, Tolmie arrived at Fort Nisqually in the spring of 1833. The preeminent naturalist, Sir William J. Hooker, had recommended him to the Hudson's Bay Company. Dr. John McLoughlin, the chief trader at Fort Vancouver and known as "the Governor," became his mentor. Tolmie's temperament, character, and intellect would serve him well as doctor and trader to the local people.

Just three months after arriving at the fort between present-day Olympia and Tacoma, Tolmie received permission for a botanizing trip to Mount Rainier. He wrote in his journal, "I am going to Mount Rainier to gather herbs of which to make medicine, part of which is to be sent to Britain and part retained in case intermittent fever should visit us—when I will prescribe for the Indians." The familiarity with the route and the prospects of good hunting enticed the Nisqually Indians Quilniash, Lashima, and Lachalet, a Puyallup Indian named Nuckalkut, and a fifth unidentified individual to serve as guides. Their 10-day trip to the park's northwest corner marked the first time that a non-native approached the mountain.

Another account involved a Yakama Indian named Sluskin; his story was translated from Sahaptin into English in a series of interviews published by Yakima-area settler Lucullus V. McWhorter. According to Sluskin, two men believed to be surveyors approached his father and some other Yakama elders about climbing Tahoma. Reluctant at first, they relented and instructed Sluskin to lead the men to the "White Mountain." Sluskin waited in camp while the men climbed from the north side, possibly the Mystic Lake area. They returned the next day to tell Sluskin that they had made the summit and saw "ice all over top, lake in center, and smoke or steam coming out all around like sweat house."

Like others before him and multitudes ever after, Mount Rainier mesmerized Second Lieutenant August V. Kautz. Of the mountain 60 miles east of Fort Steilacoom, the quartermaster and commissary officer wrote in his journal in 1857, "On a clear day [the mountain] does not look more than ten miles off…a grand and inspiring view." The mountain so hypnotized Kautz that he talked incessantly of climbing it. Despite the doubts of his fellow officers, the strong-willed Kautz made plans for a summit attempt that summer. With scant information about a route and under the prejudicial influence of the times, he wrote, "Information relating to the mountain was exceedingly meager; no white man had ever been near it, and Indians were very superstitious and afraid of it."

Kautz befriended the brilliant Nisqually Indian war strategist Leschi while he was imprisoned in the Steilacoom stockade for his leadership in the Puget Sound Indian War. Leschi suggested that Kautz take a route up the Nisqually River drainage. He probably also recommended that Kautz hire Wah-pow-e-ty, who lived in Leschi's village, to guide him. Wah-pow-e-ty agreed to guide the group; Kautz outfitted them. Each man carried an alpenstock and wore shoes with four-penny nails driven through from the inside for traction on the steep, icy slopes. No one in the party made it to the top, but Kautz ascended solo to within 400 feet of the summit. His written account helped publicize the mountain's grandeur and challenge.

The Stevens and Van Trump 1870 expedition, regarded by most historians as the first successful climb of Mount Rainier, featured Hazard Stevens, son of Governor Isaac I. Stevens. His account of the climb, like Kautz's before him, carried the prevailing attitude of the times. He wrote that, "Takhoma had never been ascended. It was a virgin peak. The superstitious fears and traditions of the Indians, as well as the dangers of the ascent, had prevented their attempting to reach the summit."

Stevens's climbing partners included Edward T. Coleman, the group's most experienced mountaineer, and Philemon Beecher (P. B.) Van Trump. Yelm homesteader James Longmire introduced them to an Indian named Sluiskin (not the same man who figured in the unnamed surveyors' climb) who the next day led them to the mountain.

Overloaded with gear, Coleman turned back, but the remaining threesome made camp at a spot that still bears Sluiskin's name. Despite the guide's warnings about the dangers above and entreaties to abandon their plan, Stevens and Van Trump set off. The two made the summit after a harrowing climb only to realize that they lacked enough daylight to make it safely back to camp. At the summit crater, they found a cave with thermal vents that emitted hot gases from the mountain's core. There they spent a damp and miserable night, alternately baking and freezing. Van Trump took a nasty fall during their descent, and once reunited with Sluiskin, the pair relied heavily on his expert knowledge of the terrain to lead them back to safety.

Despite these examples of Native American presence and expertise on and around the mountain, the belief about Indian superstitions and fears persisted well into the twentieth century. It would take years of growing archaeological evidence and an unlikely partnership to dispel the mistaken idea. University professors and graduate students, tribal elders, Mount Rainier National Park staff, and other experts would eventually work together to set the record straight.

Emerging Truth, Stubborn Bias

In the summer of 1963, there was little reason for native people in the Mount Rainier area to cooperate with Allan H. Smith. An anthropologist at Washington State University, he had few connections with local tribes, whose people were understandably suspicious of outsiders. With his colleague Richard H. Daugherty, he held a contract with Mount Rainier National Park to determine the extent and use of the mountain by Indian people. Smith planned to gather ethnographic information about their use of the mountain, which Daugherty would then use to guide an archaeological survey.

Following his interviews, Smith praised the cooperation and information he received, but noted several factors that hampered the project. The limited time available for interviews, the physical health of some informants, and their unfamiliarity with Smith all contributed to what he considered a sparse accumulation of data. But when he combined what he had learned from the 11 people on the Yakama, Nisqually, and Muckleshoot reservations with an extensive literature review, substantial new information came forward. Smith learned, for example, that Yakama, Taidnapam (present-day Cowlitz), Nisqually, Puyallup, and Muckleshoot people seasonally frequented and laid loose claim to particular areas on the mountain. While boundary lines proved arbitrary with some overlap, ridge crests generally served as approximate dividers. Smith's informants told stories of trips to the mountain in late summer to early fall where they picked huckleberries, gathered plants, and hunted elk, deer, bear, mountain goat, and other animals.

In the project's second phase, Daugherty conducted field surveys in hopes of identifying potential archaeological sites. His team located chipped stone tools in a roadcut near Bench Lake on the mountain's southern slope. They also followed up on park naturalist Terry Patton's report of a rock shelter on the eastern slope in the Fryingpan Creek drainage. These two locations marked the highest known archaeological sites in Washington at that time. The work that followed would soon change our understanding of human activity at Mount Rainier.

Anthropology graduate students and members of Daugherty's survey team David G. Rice and Charles M. Nelson began test excavations at the rock shelter in September 1964, the first study of its kind at the park. The shelter was about the size of a modern backcountry campsite, with a back wall arcing upward to form a protective roof about 16 feet overhead. There Rice and Nelson recovered chipped stone fragments indicating tool maintenance and repair, nearly one-half of which were smaller than a fingernail. They also found 13 formed tools that included knives, scrapers, and projectile points that they believed indicated connections with native people in eastern Washington. Key finds included a pumicite pipe found in a rock crevice suggesting either ceremonial or leisure activity, and bits of bone and tooth enamel from goat or sheep. Finding animal and plant remains in Pacific Northwest forests is difficult because the wet and highly acidic soils do not preserve these materials. Moisture breaks down plant tissue; acids present in coniferous forest soils dissolve unburned bones. The sheltered setting provided some protection from these natural processes, and their presence in the fire hearths helped preserve them. The recovery of the bones and tooth enamel showed that people had hunted nearby and had dressed and roasted their kills there over 1,000 years ago.

The upper valley of Fryingpan Creek, at an elevation of 5,400 feet, remains under snow from October through June, so the ancient hunters probably used the site during snow-free times between July and September. The location of their home villages remains unknown, but may have been in the Yakama lands east of the Cascades as suggested by Rice, or the lowlands west and north on the White or Green Rivers. Horse travel did not become commonplace until the 1700s, so people walked—sometimes for up to several days—from their villages to the camp.

The combined challenges of weather, terrain, and distance suggest that people had strong reasons to venture onto Tahoma's uplands. The current million-plus annual visitors to Mount Rainier come to pursue a broad variety of activities in all parts of the park, but Tahoma's precontact visitors likely came for only a few reasons. They often passed through the area as they traveled across the Cascades to visit family, trade, or for other purposes. It's possible they came for religious or spiritual practices. Most importantly, anthropologists and archaeologists believe that people came to specific locations on the mountain for the express purpose of extracting resources—plants and animals—in short supply or unavailable in the lowlands.

Most villages west of the Cascades crest sat between sea level and 1,000 feet in elevation. There, people enjoyed regular access to salmon and cedar. Deer and elk were abundant, and camas was common in lowland prairies. The dense, lower elevation forests of up to about 3,500 feet, however, held fewer valuable items than those found in the meadows or in open subalpine settings a thousand feet further up the mountain. In those areas and at other locations around the mountain, the forest edge, subalpine parks, and meadows held the greatest variety and quantity of plants and animals that people sought—at least during the snow-free summer season.

People hunted deer, elk, and bear wherever they found them, but they especially prized those animals not available in the lowlands. Hoary marmots and mountain beaver were valuable for their pelts that people sewed into blankets or robes. Mountain goat hides were treasured for their wool.

Many plants grew at mid-elevation (between about 3,000 and 5,500 feet) that benefitted native people. The long, narrow leaves of bear grass were used as part of a decorative pattern in basket making, imparting a light color to the design. Medicinal plants such as Gray's lovage were harvested to treat colds, coughs, and croup.

The subalpine meadows provided Native Americans with important food plants, too. They dug the roots of some plants and the corms of avalanche and glacier lilies. They probably prepared and ate these thickened, underground stems while living on the mountain during the short hunting and gathering season. Nuts of upland plants like whitebark pine were also harvested. There is little doubt though, that the several varieties of huckleberry that have purpled the fingers of berry pickers for innumerable generations were an essential subalpine food plant at Mount Rainier. People may bicker whether they are "huckleberries" or "blueberries," but the three varieties of *Vaccinium* were a powerful draw to Tahoma's mid-elevations. Native people favored the black huckleberry *(Vaccinium membranaceum)*. Even today, a half-gallon bag of fresh, clean berries brings top dollar. Two other varieties, oval-leaved huckleberry *(V. ovalifolium)* and dwarf blue-berry *(V. deliciosum)*, also flourished in the mountain's main berry picking areas that lay between 3,000 and 5,500 feet.

Smith and Daugherty's work and Rice's subsequent findings should have provided a new compass bearing to direct archaeological surveys on the mountain. Surely now, the stories of tribal elders supported by physical proof would unmake the myth of the region's native people as "superstitious and afraid" of Tahoma. Unfortunately, their seldom-read work moved only within small academic circles. The stone tool artifacts and bits of bone landed in a university storeroom, forgotten. Smith's report languished on a dusty shelf in the park library. The bias lingered and ignorance prevailed for another 30 years.

In the 1960s and 1970s, Mount Rainier National Park could do little to help visitors understand that they walked in the footsteps of the ancients. Park staff and visitor centers lacked the information to tell park guests that ancestors of people living nearby on all sides of the mountain had hunted and gathered plants there. The absence of place name information prevented people from connecting features like the Nisqually or Puyallup Rivers, Wapowety Cleaver (as spelled on park maps), or Sluiskin Falls to the area's original inhabitants, early guides, and travelers.

The rock shelter along Fryingpan Creek and the Bench Lake location remained the park's only documented archaeological sites for over 20 years, but the accumulation of isolated artifacts added brush strokes to its developing archaeological portrait. In his delightful book *A Year in Paradise*, original park naturalist Floyd Schmoe shared the challenge and exhilaration of his adventures as winter caretaker for the newly built Paradise Inn. A projectile point he found in the nearby Tatoosh Range in the early 1920s was one of the first artifacts collected in the park and the beginning of a series of finds that became part of Mount Rainier's archaeological record.

Two stalwart contributors to the park's collection between the 1970s and 1990s included park ranger John Dalle-Molle and long-time trails supervisor Carl Fabiani. Their finds, reports, and hand-drawn maps expanded ideas about peoples' precontact presence on the mountain. In addition to the individual artifacts that eventually totaled over two dozen, another rock shelter and a butchering site were added to the record. These two locations in the Sunrise area on the park's north side shed new light on human presence at Tahoma.

In 1990, Richard J. McClure Jr., archaeologist at the neighboring Gifford Pinchot National Forest, identified four additional sites. His work to organize the growing collection of precontact artifacts served as the foundation for the landmark work of Greg Burtchard and Stephen C. Hamilton in 1995. Working on contract with the park, the pair conducted archaeological reconnaissance on more than 3,500 acres. They added an astounding 32 finds to the park's record.

In addition to documenting new sites, Burtchard brought order and form to the increasingly robust collection of precontact artifacts. Probably his most significant contribution lay in his development of a spatial model that sought to explain why indigenous people valued mountain environments, and why they favored some landscapes over others. He reasoned that seasonally productive subalpine ecosystems stretching around the mountain answered the questions of "where" and "why" people came to Tahoma over the ages. He also developed a temporal model to explain when precontact use of the mountain began and how it changed over time. Testing and refining the models to predict where and when people frequented Mount Rainier continues to guide the park's archaeological trajectory and provides a frame of reference for evaluating and interpreting the finds.

With the park's archaeological record taking shape, its organization solid, and a model to guide fieldwork, next steps included restoring relations with local tribes and educating visitors about human activity at Mount Rainier. Burtchard reached out to tribes near the mountain, setting in motion an exchange of information and forming partnerships on a variety of projects. Training sessions for seasonal and permanent rangers soon included workshops led by local natives who knew their peoples' history at Tahoma. New interpretive programs told the stories of the original park visitors. When the new Jackson Visitor Center opened its doors, it featured cultural displays with the latest information telling the stories of indigenous people at Mount Rainier. The remodeled Sunrise Visitor Center soon followed suit. Finally, the park had gotten the story right about the precontact presence of people on the mountain.

Like Beads on a Necklace

Several factors coalesced to enable park staff to learn and tell about Native Americans at Mount Rainier. Stories from the ground played a primary role. Isolated finds like projectile points, clusters of chipped tool-stone, and tools provided clues about locations and activities. Through field reconnaissance projects, experts discovered and traced the paths of early people and formed ideas about how they used the mountain. Archaeological excavations added extensive detail to the picture of when people first came to the mountain, what they did, and how often they returned. Complementing the fieldwork, a growing relationship with local tribes helped park staff to learn their narrative. This comprehensive body of work allowed them to share with visitors the full account of the first people who came to Mount Rainier.

In his 1998 overview of Mount Rainier archaeology, Burtchard proposed nine distinct site types that had been, or would be, identified on the mountain. Among these are hunting or limited-task field camps similar to the one found along Fryingpan Creek in 1963. Small groups occupied these sites for up to a few days at a time, hunting and gathering in the vicinity. While there, they maintained and repaired tools, butchered animals, and cooked, ate, and slept. In 2001, Burtchard and his colleagues re-excavated the Fryingpan Creek camp. They exposed two charcoal-stained fire pits to find bones and teeth of mountain beaver, hoary marmot, mountain goat, deer and or elk, and bits of charred hazelnut. They also recovered projectile points, flake tools and blades, and another 1,900 chipped stone fragments. These finds all amplified the site's significance.

The distinctive feature of another field camp, at Berkeley Park on the mountain's north side, is its full cycle of stone tool manufacture. Formed after a rock avalanche crashed down Skyscraper Mountain to create a double rock shelter, fire hearths and animal remains testify to hunters' presence at the site, and to their success. The shelter's 360 artifacts represent 11 distinct stages of tool production ranging from 100 to 2,000 years old. Most of the debris suggests the final stages of on-site toolmaking, including the final touches to 13 diminutive points common to bow and arrow technology. Also found was a grooved abrader that hunters used to smooth and straighten arrow shafts.

The artifacts from the Tipsoo Lake field camp remain undated, but hold special interest because of their rarity. Researchers there found obsidian, a shiny, black, glass-like volcanic rock occurring naturally in central Oregon but not at Mount Rainier. Located between Cayuse Pass and Chinook Pass on the park's eastern edge, access was relatively easy for those coming from the east, south, or north. Plentiful water and ample protection from high winds would have made for good camping.

Butchering locations make up another type of site at Mount Rainier. Typically located at exposed and windy places that reduced the number of pesky flies, hunters used rock flakes or flake tools as cutters and scrapers to process game. A spot on the alpine tundra west of Sunrise suggests that people butchered and partially dried meat there, most likely marmot or mountain goat.

A larger type of site, residential base camps, dot Mount Rainier's parklands near the forest edge at a number of locations. Mixed age, gender, and multi-family groups probably used these sites repeatedly for extended periods during late summer and early fall to access upland plant and animal resources that they then transported back to base camps and eventually, to lowland villages. Fire hearths, small shelter depressions in the ground, and a broad mix of light and heavy tools identify these camps. One site lies adjacent to a small creek off the Kautz Creek trail leading to Indian Henry's Hunting Ground. People—possibly ancestors of members of the Nisqually Indian Tribe—used the camp repeatedly for thousands

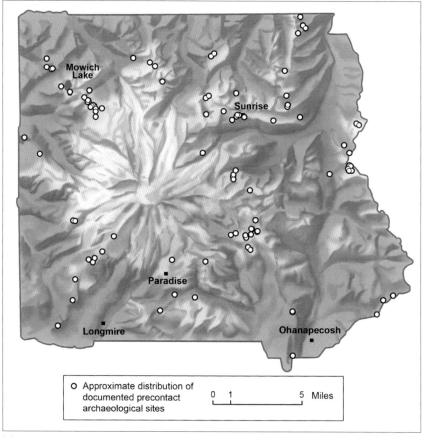

Mount Rainier's archaeological sites encircle the mountain, the majority at mid-elevation, resource-rich locations. *Greg Burtchard, Mount Rainier National Park archaeologist (retired) modified by Kirsten Wahlquist*

of years. The earliest use of the site dates to more than 7,700 years ago.

One of Mount Rainier's most extraordinary residential base camps lies in the park's northeasternmost corner, where Burtchard and colleagues conducted excavations over several field seasons. The Muckleshoot Indian Tribe lent logistical support and field assistance that enabled the team to recover nearly 20,000 stone tool artifacts. Projectile points representing both bow and arrow and the earlier atlatl technologies, scrapers, other tools, and a high density of chipped tool-stone debris comprise the bulk of the finds. Site features included a flat rock "griddle" used for cooking and multiple fire hearths with fire-cracked rock that date to about 4,200 years ago. Also found in this layer were over 300 pebble-like objects called gastroliths, the gizzard stones of grouse, a common chicken-sized game bird. Finding these stones tells the story of people roasting and eating the birds there.

The discovery and radiocarbon dating of a burned layer lying beneath volcanic ash from the Mount Mazama eruption nearly 7,700 years ago unlocked the riddle of how far back in time this site was occupied. There the team found a tool fragment, some tool-stone flakes, and a gizzard stone. Analysis of the charcoal samples associated with the layer and its contents aged them at between 7,520 and 10,240 years old. Burtchard believes that a good estimate for earliest use of the area is between 8,000 and 9,000 years ago, although more study could move the date back even further. The site's location within traditional Muckleshoot areas makes it reasonable to infer that ancestors of present-day tribal members probably used it.

One way to appreciate the value of Mount Rainier's growing archaeological record is to compare it to the precontact currency system of Native Americans in the region—the dentalium shell bead necklace. As shells were added, the value of the necklace increased.

A little more than 50 years ago, Mount Rainier's precontact record consisted of a small collection of isolated finds enveloped in a cloud of misunderstanding about the presence of Native Americans on the mountain. The sustained cooperation of local Indian tribes, park staff, and other experts finally prevailed in dispelling the mistaken idea that indigenous people avoided mountainous terrain. For over 9,000 years at more than 100 locations, native people have hunted marmots, mountain goats, and other game. They have gathered huckleberries, bear grass, and other plants. Tahoma's archaeology now places people on all sides of the mountain, encircling it like shell beads on a necklace, becoming ever more valuable as new finds are added with each passing year.

3

The Nisqually River, From Glacier to Sound

Mount Rainier from the Nisqually River, near the Cougar Rock Campground. The bare riverbank and scattered logs illustrate the river's erosive power. *Drawing by Lucia Harrison*

GLACIERS: WHERE RIVERS ARE BORN

In his 1894 letter to Congress urging designation of Mount Rainier as a national park, geologist Bailey Willis described the mountain as "an arctic island in a temperate zone." He wrote that the glaciers "are themselves magnificent, and with them survives a colony of arctic animals and plants which cannot exist in the temperate climate of the less lofty mountains. These arctic forms are as effectually isolated as shipwrecked sailors on an island in mid-ocean."

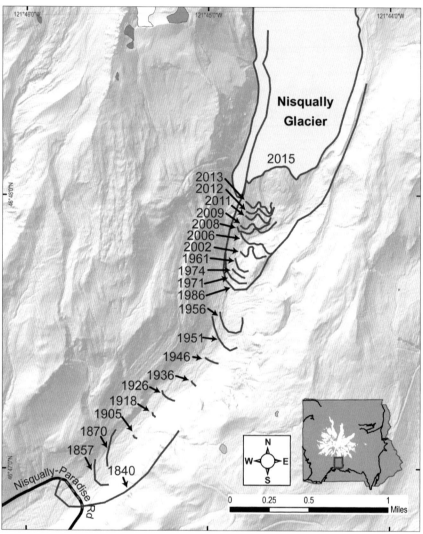

The terminus position of the Nisqually Glacier since 1840. Notice that it advanced from 1961 to 1971, retreated in 1974, and advanced again in 1986. All measurements before and since have indicated glacial retreat. *Modified from Beason et al. (2017) by Kirsten Wahlquist*

While Willis's words made for great imagery, they separated the mountain from the Puget Sound in much the same way that lines on a map delineate ownership. But rivers are powerful connectors, and emerging from the Nisqually Glacier, the Nisqually River is one of the region's mightiest. It churns and glides 78 river miles as it stitches Mount Rainier's alpine heights to the Salish Sea lowlands. The best way to treat its natural history properly is to tell the watershed's epic story from glacier to estuary.

Mount Rainier's glaciers are magnificent, but travelling upon them is a dangerous proposition for even the most seasoned mountaineer. It demands physical stamina, mental focus, and an ability to interpret a complex set of external cues. An axiom among locals that "the mountain makes its own weather" derives from abundantly available moisture that combines with the environmental gradients of temperature and elevation. This potent mix creates fierce, sudden storms that can materialize with little or no warning. Clouds and snow sometimes merge in a disorienting whiteout, leaving no visible horizon and no points of reference. Unprepared or inexperienced travelers can easily become lost under such circumstances.

Danger lurks on a glacier, even on the clearest days when climbers marvel at the sprawling Cascade spine stretching north to British Columbia and south to Oregon and beyond. Crevasses yawn with menace. Snow bridges collapse without warning. Avalanches and rockfalls trigger spontaneously. New hazards appear during extended periods of hot weather that cause climbing routes to deteriorate.

The Nisqually Glacier is the mountain's biggest south-facing glacier and at 1.62 square miles, its sixth largest. Relatively easy access has enabled researchers to compile a 150-year record of its many advances and retreats. Among the most studied glaciers in the western hemisphere, the first photos of its terminus date to 1884. After glacial measurements began in 1905, scientists closely followed its one-mile flight up valley in the first half of the twentieth century. Overall, it has retreated about 2.5 miles since 1840, its terminus moving from 3,800 to 5,500 feet in elevation. In a recent eight-year period, it retreated up the mountainside nearly two-tenths of a mile, more than three feet every 10 days.

As Mount Adams and Mount Hood dazzle to the south, a few other trained volunteers and I help park geologists measure ice velocity, a warning sign of glacial outburst floods. We work on the glacier's lower section out of harm's way, safe from crevasses and rockfall, but remain on high alert. There is movement everywhere, some of it undetectable. We ride, for example, on a river of ice that runs to over 400 feet thick in places, flowing over the ground beneath it at a rate of several inches to several feet per day. At one spot we hear water running under our feet. From another we watch a distant rockfall rumble down a lateral moraine, hurtling debris skyward. A volunteer takes a misstep and tumbles, a lacerated elbow a painful reminder to stay focused at all times. Nearby, the river gushes forth from the glacier's terminus in a roiling fury of clambering bedload, rocks banging against the river's bed. In a fit of splash and spray, the Nisqually River sees its first daylight and rushes off, hell-bent for the lowlands like a weary backpacker craving a cheeseburger.

SUBALPINE MEADOWS AND THE MOUNTAIN HEMLOCK ZONE

A scant mile southeast of the Nisqually Glacier's terminus, Paradise sits on the mountain's broad shoulder. It got its name in the late 1800s from James and Virinda Longmire, who held a mining claim and operated a hotel in the valley below. Virinda Longmire described the valley as having "lawns of green grass covered with beautiful alpine flowers of varied hues...paradise." The Longmires weren't the first to visit the meadows, of course. Given the proximity of ancestral Nisqually villages, it is likely that indigenous people went there seasonally for generations.

John Muir, the revered naturalist and conservationist, visited the Pacific Northwest in 1888. At age 50, he had not planned to scale Mount Rainier. "I did not mean to climb it," he wrote his wife Louie, "but got excited and was soon on top." Guided by veteran mountaineer P. B. Van Trump and Major Edward Ingraham, each of them namesakes of geographic features within the park, the party of nine made their approach from the Nisqually side. Muir selected the group's campsite at 10,100 feet. Camp Muir still bears his name as a base camp for climbers and a popular destination for day hikers.

Struck by the beauty of the flower-laden meadows, Muir penned the words now inscribed on the stone staircase leading from the Jackson Visitor Center to the network of Paradise hiking trails: "The most luxuriant and the most extravagantly beautiful of all the alpine gardens I ever beheld in all my mountain-top wanderings." One of the park's first ecological studies called the meadows "among the most famous natural flower gardens in the world." Another report found that nearly one-half of the park's visitors came to view the wildflower meadows. The lure continues. The acres of lupine blue, lousewort yellow, paintbrush scarlet, and valerian cream cast such a spell that *Wildflower Wonders: The 50 Best Wildflower Sites in the World* advises readers that, "if you only visit one site to see North American wildflowers, then this should probably be it...it is the sheer spectacle that ranks highest."

The park's subalpine meadows border its uppermost forest, the mountain hemlock zone, which begins at about 4,000 feet and links the

The Paradise subalpine meadows at their August peak. How many species can you identify? © *Keith Lazelle Nature Photography*

Subalpine lupine and one of the park's handful of paintbrush species, two icons of the Paradise flower fields. ©*Keith Lazelle Nature Photography*

closed-canopy forests to the meadows. Some call mountain hemlock "the star of the high country," because needles cover its branches on all sides, giving a star-like appearance. Although mountain hemlock is often found where snow persists until midsummer—like it does at Paradise—it keeps growing at near-freezing temperatures if its roots remain thawed. Other trees of this zone include yellow cedar, silver fir, and the ever-present and increasing subalpine fir, all of which dot the Paradise parklands. Subalpine fir's spire shape allows it to shed snow easily. A firm handshake with a branch—being careful to leave the needles on the tree—imparts a pleasantly fresh citrus smell, my favorite "Tahoma Aroma."

Where the continuous forest line of the mountain hemlock zone tapers off at about 5,300 feet, the subalpine meadows begin and extend up slope to 6,600 feet on the south and west sides of the mountain. Drier conditions on the north side in the Sunrise area allow the meadows to stretch from approximately 5,700 feet to 6,900 feet in elevation. Encircling the mountain at mid-elevation, Mount Rainier's meadows cover nearly 40 square miles. In summer, resplendent wildflowers spangle the mountain's emerald apron. Autumn snowfall transforms the meadows into a gleaming white winter cloak.

Glaciers covered the meadows in ancient times, most recently during the Evans Creek Glaciation between 22,000 and 15,000 years ago. When the climate warmed, the glaciers retreated. Pioneer plants gradually began eking out a tenuous existence. The gravelly, rocky soils comprised mostly of glacial till or volcanic tephra provided few nutrients. Annual snowpack depth and the length of time that snow remained on the ground affected meadow vegetation, too.

Of the major plant community types that characterize the Mount Rainier subalpine zone, the main attraction for visitors is the lush-herbaceous community, which thrives at Paradise. Many of these plants grow to a height of 18 inches or more, and are easily identified with the help of a wildflower identification card, field guide, or mobile phone app. The lush-herbaceous community also occurs on the mountain's north side at Sunrise, but there it has less biodiversity and a greater abundance of the less showy mountain bunchgrass. Park regulations remind visitors to "Leave No Trace" by picking no flowers, staying on trails, and keeping off the fragile subalpine vegetation.

Some of the common and conspicuous species include the bluish, pea-like flowers of subalpine lupine and the white to pale-pink, dense flower heads of Sitka valerian. High country hikers often notice valerian's pungent, sour odor after autumn's first hard frost. Another key plant, green false hellebore (*Veratrim viride*), grows in moist areas. Its flowers are nondescript but its height, up to nearly six feet tall, makes it a literal standout. Also called Indian hellebore or corn lily, it is one of the most poisonous plants in the Pacific Northwest and a respected and powerful medicine among indigenous people and herbalists. Used in the proper manner and amount, hellebore is used to treat toothaches, sprains, bruises, colds, and other ailments.

Western anemone or pasqueflower (*Anemone occidentalis*) is another common poisonous subalpine plant. Its shaggy, fruiting seed head follows a handsome, early season, creamy white flower. Other common names like mouse-on-a-stick or mop-top capture its Dr. Seuss-like appearance.

POLLINATORS

Observant subalpine hikers often notice the buzz and whir of aerial traffic as pollinators make their summer rounds. Although some plants self-pollinate and wind helps to pollinate others, the major role belongs to airborne species. Up to five species of bumblebees pollinate members of the figwort family, including monkeyflowers, penstemons, louseworts, and paintbrushes. Bumblebees and their close relatives the honeybees seek the carbohydrate-rich, life-sustaining nectar found within each plant's corolla. As the bees search the meadow flowers for nectar, they collect the fine, powdery pollen that contains the male cells needed for plant reproduction. Aided by leg combs and brushes, they place the pollen in rear-leg storage baskets, then deposit it onto the flower's stigma, the tip of its reproductive organ, where the pollen germinates.

The western bumblebee, once an important subalpine pollinator, experienced a drastic die-off in the late 1990s that experts attribute to a fungal parasite transmitted by commercially raised bees. The status of western bumblebees remains precarious, but recent sightings indicate that some bees with fungus-resistant genes may have survived and that their offspring are staging a modest comeback.

Adaptive strategies equip bumblebees for climes where summer lasts for weeks, not months. The use of rodent burrows as pre-made nest sites, proportionately large, furry bodies, and the ability to maintain body temperature all contribute to subalpine success. Unlike their honeybee cousins, bumblebees do not produce enough honey to feed either humans or bears. Instead they make just enough to sustain their larval broods and the hive. Only the hibernating queens survive the winter, which helps preserve food supplies.

Other subalpine pollinators include over a dozen species of Syrphid flies. People may mistakenly call them sweat bees, but they are actually a group called the hoverflies. They can outnumber bumblebees at Mount Rainier. A third group of pollinators, the Muscids, includes houseflies. They remain active when temperatures dip into the mid-40s, conditions that render bumblebees and Syrphid flies inactive.

Subalpine parkland pollinators also include the rufous and calliope hummingbirds. The rufous winters from southern California into Mexico and Texas and occurs with increasing regularity in the southeastern United States. It logs up to 8,000 miles per year traveling up the Pacific Coast to Alaska, then looping inland and south. Rufous hummingbirds begin arriving in the Pacific Northwest lowlands in late February and early March just as red-flowering currant and salmonberry begin to bloom. Rufous are aggressive in defending their feeding territories, not only against each other but also against larger birds and even people. Soon after breeding, the males move up to the subalpine meadows, where the abundant flowers hold the precious nectar that the birds will convert into body fat and later burn as fuel for their southbound migratory flights. Females and juveniles depart after the males in late summer. They migrate not in flocks but individually, stopping occasionally to feed in order to replenish their energy reserves.

The calliope hummingbird is the smallest breeding bird in North America, averaging about the weight of a single U.S. penny. One good look at the male's iridescent magenta gorget in bright sunlight explains the meaning of part of its scientific name, *Stellula*, which means "little star." Its path arcs northward to the central British Columbia mountains and then south to its wintering grounds in Mexico. The mountain-dwelling calliope typically breeds between 4,000 and 13,000 feet in elevation.

Rufous and calliope hummingbirds are sympatric at Mount Rainier, meaning that they occupy the same areas, namely the subalpine meadows. They have strong spatial memories and remember those flowers drained of nectar, those with nectar remaining, and those not yet frequented. The rufous is the more pugnacious of the two, so researchers were curious to know how the calliope feeds successfully in the same habitat. They found that calliopes escape detection by rufous hummers by flying exceptionally low into feeding areas, "flying below the radar" so to speak. They often perch on the ground, and tend to feed on flowers closer to the ground. A faster feeding rate allows them to drink as much nectar as quickly as possible before the rufous chases them away.

The two birds compete against each other to pollinate various species of paintbrush, one of the iconic symbols of the Pacific Northwest mountain summers. Paintbrush flowers are green and inconspicuous, but the eye-catching orange, magenta, and scarlet bracts attract hummers. Anyone wearing a red cap or bandana may learn of the hummingbird's preference for red as the tiny powerhouses boldly investigate. They are the primary paintbrush pollinators at Mount Rainier; the process is straightforward. As the hummer probes for the energy-giving nectar, it inserts its beak into the flower's corolla. Pollen collects on the bird's forehead. It then flies on to the next flower to continue searching, and deposits the pollen to finish its job.

The relationship between subalpine plants and the animals that interact with them spark great interest among researchers. Animals worldwide aid in the reproduction of most flowering plants, with the bulk of high country activity occurring during the one- to two-month long growing season set in motion by the disappearance of snow. Once the snow melts out, the annual cycle that includes bud and flower formation, fruit maturation, and seed dispersal begins. This attracts those organisms that transfer pollen and benefit from the nectar, fruit, and seeds produced. Scientists call this seasonal timing of biological events phenology. Climate affects phenology, with warmer temperatures causing the majority of plants to advance their growth and reproductive sequences. When snow melts earlier in the season, plants begin their growing cycles sooner. This creates advantages for some plants and challenges for others, and can cause timing mismatches for the organisms that depend upon them. A hummingbird or insect, for example, might arrive to find that the nectar it seeks is no longer available because the plant flowered earlier than usual.

Researchers from the University of Washington studied 48 plant species at Paradise to examine the relationship between phenology and climate. They found that snowmelt, temperature, and soil moisture exerted strong influence on flower timing and duration, modifying the community's makeup. The interactions between species shifted, as did the number of individual plants per species and the sites they occupied. The researchers also noted changes in the overlap of flowering between species, labeling the sum of these events as "community reassembly." Reassembly could reshape the interactions between plant species, possibly including competition for pollinators, which in turn could trigger population increases or declines.

A stunning discovery came during one unseasonably warm, dry summer when snow began melting out 58 days earlier than in any of the previous five years of the study. The early snowmelt, warmer temperatures, and drier soils gave a preview of the extreme conditions predicted for subalpine plant communities near the end of this century. All of the species observed flowered earlier that year and over half increased the length of time that their flowers persisted, some by as much as 15 days. Other species experienced shorter than usual flower duration periods. A mash-up of previously unseen patterns resulted in community reassembly, including the simultaneous flowering of some species that normally bloom weeks apart from each other.

It is unclear if that event can reliably predict what these communities will look like later in this century. Fluctuations in global CO_2 emissions or other circumstances may alter future conditions. If the full force of climate change continues, however, models indicate that the recent scenario provided an analogous example of what lies ahead for subalpine plant communities at Mount Rainier.

Besides phenological changes and plant community reassembly, climate has been gradually remaking the mountain's mid-elevation ecosystem for decades. Subalpine fir, one of the dominant trees in the mountain hemlock upper forest zone, began growing in the Paradise meadows in substantial numbers during a warmer and drier period in the 1930s. Knowing that the vibrantly colored flower fields attracted throngs of visitors each August, park staff worried whether the encroaching trees would eventually crowd out the wildflowers altogether. The wording of the 1916 Organic Act that created the National Park Service complicated the question of whether to remove the trees in order to preserve the rainbow of color. According to the act, the service's mission is to "conserve the scenery and the natural and historic objects" while "providing for the enjoyment of the same," yet "leave them unimpaired for future generations." The park chose to remove the trees and did so until 1979. A policy change then allowed the trees to stay, and they continue to crowd their way into the meadows.

Research showed that the trees' incursion correlated with variations in snowpack thickness. Trees establish more easily during warm and dry periods on the mountain's west side where the snowpack lies higher in elevation. As Earth continues warming, the firs will likely continue to displace the wildflowers at Paradise.

Climate-caused modifications to subalpine ecosystems at Mount Rainier mirror some of the alterations to similar plant communities globally. Here and elsewhere, these wholesale changes figure to benefit some species while handicapping others. Local researchers remain uncertain about those that will emerge as winners and which as losers. With continued study, they hope to track and understand the long-term effects of climate on subalpine plant communities.

Loving the Meadows to Death

In *A Sand County Almanac*, Aldo Leopold wrote, "Man always kills the things he loves, and so we pioneers have killed our wilderness." Leopold's dictum is especially true for the meadows at Mount Rainier. For over a hundred years, visitors have swarmed Paradise on foot, on horseback, and by motor vehicle. The heaviest impact occurs on weekends during the summer months. Prior to becoming a national park, local newspapers reported overhunting and human-caused fires. In 1907, Mount Rainier became the first national park to admit automobiles. The first car crept up to Paradise in 1912 and within a few years, drivers parked indiscriminately in the meadows. Other contributors to meadow damage included a tent camp business, a downhill skiing operation that lasted for 40 years, and even a short-lived nine-hole golf course. The horseback riding concession created some of the most severe and longest lasting impacts, leaving trenches two feet deep and over 10 feet wide that lingered for nearly 30 years after the last pony left Paradise.

The meteoric rise of backcountry hiking in the 1970s added to the widespread degradation of the subalpine meadows. Record numbers took to the trails at Olympic and Mount Rainier National Parks, with each boot print holding the potential to release a choking cloud of dust onto nearby plants. High volumes of foot traffic can trample vegetation and damage sensitive areas in less than two weeks. Recovery time at low elevations takes approximately the same amount of time as that taken to inflict the damage, but at higher elevations, it can take from 10 to 1,000 times longer. In places where the growing season is short and the conditions demanding, restoration takes years or decades.

After a century of intense human presence at Paradise, the impact was considerable. More than 900 shortcuts known as social trails had left jagged, bare-ground scars between designated trails. Improvised and unregulated campsites and picnic areas blemished the parklands, eroding an otherwise lush landscape. Overuse especially affected the heath meadows, which range from around 6,900 to 7,800 feet in elevation and require at least 200 years to progress from bare ground to established, mature communities. In contrast to the lowland old growth forests that may live up to 1,200 years, heath meadows can endure for over 7,000 years. They are more sensitive, however, to trampling than any other plant community at Mount Rainier. As Leopold had written decades earlier, our obsession with Paradise was killing it. We were loving it to death.

Ironically, the solution was human intervention. The comprehensive meadow restoration program that began in the mid-1980s was not the first work of its kind at Mount Rainier. Beginning in 1930, crews had revegetated subalpine areas at Paradise. Later in the decade, Civilian Conservation Corps workers moved transplants from a spot destined to become a roadbed to an exposed location in need of plant cover. In the late 1950s, asphalt trails were the best option for those badly eroded in the lower meadows. Elsewhere, a landmark project at North Cascades National Park had proven that greenhouse-propagated plants could take hold and thrive in the high country. That success became the blueprint for restoration plans built around greenhouses at Olympic and Mount Rainier, providing the plant stock that mends subalpine meadows.

Leading-edge techniques pioneered at the three national parks begin with collecting seeds near the impacted site in order to preserve genetic integrity. Close attention to exposure, slope, and elevation of the collected seeds helps increase the probability of survival as fledgling plants. Greenhouse staff carefully store and germinate the seeds and then raise the young plants over the winter and spring months. In conjunction with its greenhouse program, Mount Rainier's Paradise project soon expanded to other areas in the park. First, staff documented impacted areas. Then they developed, implemented, and evaluated restoration techniques. Many of those techniques continue today as the standard on public lands throughout the Pacific Northwest.

The labor-intensive restoration process begins with loosening and breaking up heavily compacted soil. Fill material is often added. Workers install cut logs or rock slabs as stabilization bars to help control erosion. They may also lay down mats made of wood shavings or other absorbent materials to regulate moisture and soil temperatures in preparation for the late summer and early fall plantings. Crews then transport the greenhouse stock to the meadows and hand-plant them by the thousands. The transplants jump-start rehabilitation projects and signal park visitors that restoration is underway and to avoid the area. Workers clump plants together, which creates micro-conditions for protection from extremes of sun and wind and improves the odds of survival. Plants clumped together also have a natural look that helps change visitor traffic patterns, allowing denuded ground to heal more quickly. The transplants provide cover for other seedlings and soon contribute seed for other plants to develop. Seeding, layering, and allowing areas to revegetate naturally also play roles of varying importance in meadow restoration projects.

Mount Rainier's staff studied visitor use patterns to understand non-compliance with signs and fences intended to protect sensitive areas. They experimented with issuing citations and cordoning off areas with yellow nylon rope. Eventually, park staff discovered that the presence of uniformed personnel was the most effective way to limit meadow trampling. This gave rise to the Meadow Rover program, in which trained volunteers patrol the meadows during the summer months to interact with visitors.

The Mount Rainier meadow restoration project is an ongoing success story of human intervention built on research, innovation, and commitment. It has taken dozens of years and thousands of hours of manual labor, much of it done by volunteers. Whether collecting seeds, transplanting, or being a Meadow Rover, abundant opportunities allow visitors to help steward a national park and deepen the connection between the place and its people.

People of the Grass

People have lived in the Nisqually drainage for thousands of years. Experts remain unclear on the origin of people in the Pacific Northwest. Nisqually elders have remarked, "It really doesn't matter, we have always been here and we are still here." One version of how they came to the land is from tribal historian Cecelia Svinth Carpenter. She wrote that, "A Nisqually tribal legend relates how people once lived near the sun in the lands of Central America, where the great ice sheets could not reach them." Over an immense span of time, the people moved northward through the Great Basin to present-day eastern Washington. They eventually crossed over the Cascades to settle in the Nisqually watershed. Here they became Salmon People. They called the river Squalli after the tall, flowing grasses on the nearby prairies. In Lushootseed, they were the *dxʷsqʷaliʔabš*, calling themselves Squalli-Absch, "people of the river, people of the grass."

Of the dozen-plus year-round Nisqually villages, none were within current park boundaries, but stories tell of people settling near Ta-co-bet, their name for Mount Rainier. Squaitz village stood on Skate Creek near Bear Prairie, just a few miles south of present-day Longmire. Lah-al-thu village sat further downriver on the north bank of the Nisqually near today's town of Elbe. Lieutenant Kautz noted this village in his journal during his 1857 summit attempt. Northern Pacific Railroad surveyors mapped it 10 years later, and Hazard Stevens mentioned it in his account of his expedition to climb the mountain.

Just below the Alder and LaGrande dams on the Nisqually stood the village of *Sákwiabc*, which may have been the largest Nisqually village in the 1850s.

Another notable village, known as Mashel, Me-schal, Meshal, and other names, sat on a bluff a few miles downriver, where the Mashel River empties into the Nisqually. People there spoke both the Lushoo-tseed language of the Coast Salish people and the Sahaptin language of the people from east of the Cascades, suggesting some migration of Sahaptin-speaking people into the Nisqually watershed. Mashel became the home of Soo-too-lick, known widely as Indian Henry. A legendary figure, he became a prosperous guide, hunter, and farmer who helped plot the town of Eatonville. Tales of Soo-too-lick's prospecting success in the park's southwest corner are probably fable, but his larger-than-life reputation was preserved in the name of the broad subalpine meadow called Indian Henry's Hunting Ground. Satulick Mountain also remembers him.

Wah-pow-e-ty, the guide for Kautz's summit attempt, also came from this village. Mashel's greatest historical figures, however, were brothers and acclaimed leaders Quiemuth and Leschi. Dissatisfied with the proposed location of the newly designated Nisqually reservation, the charismatic and eloquent Leschi refused to move there. He organized an armed resistance that was the genesis of the Puget Sound Indian Wars of 1855–56. The territorial government soon arrested and charged him with the murder of a militia colonel. After two trials, Leschi was found guilty of murder and sentenced to death. His appeal failed, and he was hung on February 19, 1858. William Tolmie, Lieutenant Kautz, and Seattle pioneer Ezra Meeker each followed the case closely and found the proceedings badly tainted under suspicious circumstances. Meeker's *The Tragedy of Leschi* remains the definitive work detailing the injustice.

In 2004, the Washington State Legislature convened an educational, historical court proceeding to reexamine Leschi's verdict. While not legally binding, the panel ruled that he had acted as a wartime combatant and exonerated him of murder.

About 10 miles upriver from where the Nisqually rolls quietly into the Sound, the ancient village of *Yo'xwálscabc* holds dual significance. Tribal historians believe that the village thrived at the confluence of the Nisqually and Muck Creek (also called Dog Salmon Creek) for thousands of years. It became part of the Nisqually Indian Reservation and maintained its identity longer than most other Nisqually villages. The creek there still supports runs of chum (also called dog salmon) and Puget Sound steelhead.

Yo'xwálscabc was also the birthplace and home of Qu-lash-qud. Born around 1879, he was a living link between nineteenth and twentieth century Coast Salish life. He verified the accuracy of ancient Nisqually village locations as recorded in 1940, and made valuable contributions to Dr. Allan Smith's work in 1963 by providing a storehouse of information about Nisqually and Puyallup lifeways. He vividly remembered his days as a youth in the Nisqually watershed, calling it "Paradise." In Qu-lash-qud's paradise, "There was plenty of everything." Wild carrots, potatoes, onions, roots, and other bulbs sat ripe for the digging. Game animals and fish of many kinds were plentiful.

When Qu-lash-qud grew to adulthood, he passed his vast knowledge onto his children. Qu-lash-qud's English name was Willie Frank, and his principal student was his first-born son, Billy Frank Jr.

PARTNERING FOR THE COMMON GOOD

Ohop Creek is one of the Nisqually's two salmon-bearing tributaries, emptying into the river a few miles from the town of Eatonville. In the 1800s, the area's fur trappers decimated beaver populations. As a result, beaver dams failed and the resulting surges of water carved the Ohop Valley floor into a mosaic of freshwater wetlands in a deep forest. The newly created salmon habitat eventually supported all five species of Pacific salmon. But when European American homesteaders began arriving in the 1880s, they had farming, not fishing, on their minds.

Shortly after settling in the valley, the newcomers began re-routing the upper portion of Ohop Creek to improve drainage and enhance farming. Over the next 25 years, they straightened the twisting, braided creek into a single, fast-flowing ditch. The new drainage pattern helped their dairy farms prosper, but spelled disaster for the fish that over-wintered in off-channel habitat and spawned in the creek. The new ditch lacked the streamside vegetation that once shaded and cooled the creek's waters that made ideal spawning conditions. Water temperatures increased and the salmon stopped returning.

Within a few generations, the degraded habitat of Ohop Creek was a microcosm of conditions elsewhere in the Nisqually watershed. An influx of people created a housing boom and new pressures on the land. Timber and agricultural operations did well at the expense of deteriorating and lost habitat. The salmon were in trouble. By the 1970s, many runs had dropped to record lows; some teetered on the edge of extinction. Of the several responses to the gathering catastrophe, one by the Washington State Legislature called for the creation of a Nisqually River management plan. Special interest groups normally at loggerheads with each other were set with the task of finding common ground on a host of issues concerning the Nisqually River. Around a table sat representatives of timber, farming, conservation and environmental groups, private landowners, state and federal agencies, the U.S. Army, and the Nisqually Indian Tribe. One night at a meeting amidst an atmosphere of mistrust and bickering self-interest, a Nisqually tribal leader rose to speak. Billy Frank Jr., son of Qu-lash-qud, told the group that the tribe wanted the forest products industry and the family farms to continue operating, but that the salmon needed protection, too. He told the group that if they worked together they could find solutions together.

Frank's powerful words united the task force and helped forge an unlikely alliance that advocated for balanced stewardship of economic, natural, and cultural resources. It is likely that no one realized that the precedent of these improbable partners working for the common good would become the legacy of the Nisqually watershed. The legislature adopted the group's Nisqually River Management Plan, which called for the establishment of the Nisqually River Council. The council's role in protecting and restoring the river would become an international model for its ambitious watershed protection plan.

On a damp and monochrome February morning, I helped students from The Evergreen State College in Olympia prepare to assist Eatonville Elementary School students sample and test the waters of Ohop Creek. The site is a short bus ride from their school and soon, 30 fourth-graders would enthusiastically test water samples for pH, temperature, and other water quality parameters. The Nisqually River Education Project oversees the work, connecting students in grades 4 to 12 with the creeks and streams near their schools.

The creek that the fourth graders energetically invaded on that gray morning looked far different from the ditch settlers had dug over 100 years before. In fact, it looked more like the Ohop Creek of 200 years ago. Instead of a straightened, sterile ditch, this section burbles in a series of broad and lazy meanders reminiscent of its ancestral path. The Nisqually Tribe and its collaborators had just spent several years and millions of dollars to convert a half-mile of the Ohop Creek ditch into a 1.1-mile stretch of wandering creek. First, giant earthmovers chugged for weeks to carve a new channel on the 250-acre tract. Bolstered by their success on the Mashel River at nearby Smallwood Park, workers then installed 42 engineered logjams (ELJs) to replicate natural stream processes, reduce erosion hazards, and enhance salmon habitat. Crews use heavy equipment to install groups of mature logs in specific locations on the waterway. The logjams slow the rate of water flow and allow sediments, sand, and gravel—essential materials for spawning salmon—to accumulate. These features create what biologists call habitat structural complexity, a great improvement over the outdated practice of removing woody debris from rivers that increased flow rates. The ELJs also give salmon a place to search for food, rest, or hide from predators.

With the ELJs in place, volunteers helped plant 86,000 native shrubs and trees. Subsequent work lengthened the restored meander to 2.4 miles and increased the number of native plantings to over 150,000. Many plants line the creek, once again providing the shade and cooler temperatures the fish need for spawning, and lending cover for birds and other wildlife, too.

The installation of ELJs on the Mashel River at Smallwood Park nearly tripled the numbers of juvenile coho salmon, and results on the Ohop were equally encouraging. The creek runs cooler and slower now as juvenile coho wait out winter floods in the off-channel habitat. Data from a tagging project will help drive future restoration efforts. Steelhead and Chinook salmon, each listed as threatened under the U.S. Endangered Species Act, are also staging comebacks. The Chinook depends on the Ohop and the Mashel as the Nisqually's only tributaries capable of sustaining its populations. Looking ahead, similar projects aim to restore and enhance other salmon-bearing habitat on the Nisqually and its tributaries.

Technicians check an engineered logjam downstream of Eatonville's Smallwood Park near the junction with the Little Mashel River. *Northwest Indian Fisheries Commission*

Camas dominates a Joint Base Lewis-McChord prairie in a spectacular spring bloom. The dainty white flowers in the foreground are chickweed; biscuitroot (*Lomatium triternatum*), yellow, is mid-ground. The white stakes identify important environmental resources on the military installation. *Adrian Wolf/Center for Natural Lands Management*

Many individuals and organizations contributed to the success of the Nisqually's restoration projects. In the Smallwood Park and Ohop Creek projects alone, an array of partners contributed financial or staff support, technical expertise, equipment, or facilities. Thousands of volunteer hours provided much of the manual labor. The South Puget Sound Salmon Enhancement Group, the Pierce County Stream Team, and other organizations joined with the Nisqually Tribe to undertake and complete this landmark work.

A key contributor to the watershed's restoration is the Nisqually Land Trust. Projects like those on the Mashel River and in the Ohop Valley begin when landowners wish to sell their land directly to the trust or create conservation easements. The trust stewards more than 7,000 acres, which it manages in 13 habitat blocks, or "protected areas." These chunks of land in turn connect shoreline properties safeguarded by local, state, and federal agencies and the Nisqually Indian Tribe. "Small watershed, big ideas," is the trust's motto, and it holds true. When stakeholders created the Nisqually River Management Plan, just 3 percent of the river's 42 miles of salmon-producing shoreline benefitted from permanent conservation status. The plan called for 90 percent protection—a bold vision, and outside the Nisqually, one widely considered unachievable. But today, 77 percent of that shoreline is in safekeeping, and the total grows every year.

The trust works on other projects, too. The Mount Rainier Gateway protects over 2,500 acres of varied habitats near Mount Rainier and around the town of Ashford. The Nisqually Community Forest—totaling 1,920 acres and Washington's largest community-owned forest—includes critical habitat for Chinook salmon and steelhead trout on Busy Wild Creek, the headwaters of the Mashel River. The forest serves local communities by providing forest products, job opportunities, and recreational, educational, and environmental benefits.

PRAIRIES FOUND AND LOST

Whatever they expected to find in the Salish Sea lowlands, the first European explorers and trappers encountered the unexpected. Instead of the endless, uninhabited wilderness they imagined, they found long-standing cultures built upon complex social systems with intricate patterns of subsistence and commerce. Along with expansive forests, they found other varied landscapes.

British Captain George Vancouver and his crew entered Puget Sound in the spring of 1792 to find huge, treeless tracts of land reminiscent of the parklands in their native England. Vancouver's journal told of "verdant, open places…these beautiful lawns," and that, "the surrounding country… presented a delightful prospect, consisting chiefly of spacious meadows." The largest prairies in Vancouver's time laid along the traveler's route between the Columbia River and the Puget Sound area, just east of today's Interstate 5.

Newcomers described the lands variously as prairies, grasslands, meadows, and savannas. In the spring of 1806 along the lower Columbia River, Lewis and Clark's Corps of Discovery found prairies awash in deep blue camas blooms and others full of wild onion. David Douglas, the intrepid Scottish botanist and namesake of over 80 species of plants and animals, also found the prairies captivating. Having wandered over 7,000 miles west of the Rockies, he noted the "extensive natural meadows…a profusion of flowering plants."

The prairie-oak ecosystem once stretched from the southern British Columbia interior valleys to central California. Communities of grasses, flowering plants, shrubs, and Garry oak established themselves during a warm and dry climatic period between 11,000 and 7,250 years ago. The Pacific Northwest prairies spanned over 300 miles from Vancouver Island to the Willamette Valley in Oregon, a grassy patchwork growing within a wide range of soil and hydrologic conditions. The soils often run gravelly and shallow, draining well but drought-prone. The north-south band of intermittent prairies in the lower Nisqually watershed lay upon gravelly terraces of glacial outwash left behind by the Vashon Stade of the Cordilleran Ice Sheet. These open habitats consist of perennial bunchgrasses and forbs with little or no woody vegetation. Herbaceous flowering plants outnumber native grass species like Roemer's fescue and wild blue rye.

In a region well suited to deep forests, these ecosystems thrived because of fire's integral role in the prairie's ecological processes. Evidence suggests that fires resulting from both natural causes and those set by people helped maintain them, preventing the forest from encroaching upon the grasslands. Seasonal and intentional low-intensity fires kept Douglas fir and other early successional species at bay and preserved the open landscapes needed by Garry oak, camas, and other plants of value to native people.

On the Nisqually prairies and elsewhere, Native Americans gathered Garry oak acorns, which they ate raw or baked in pit ovens. They also boiled them, placed them in baskets, and buried them in lake mud. When dug up later, the acorns shelled easily.

Indigenous people especially prized camas as a food plant. Exchanged widely, it ranked second only to salmon as a trade item. Entire families dug the larger bulbs in the springtime, leaving the smaller ones for later harvest. Then they built a fire in a pit a few feet deep and tended the leaf-wrapped bulbs for hours or days. If not fully cooked, they could cause gastric distress. Captain Meriwether Lewis wrote, "When eaten in large quantities they occasion bowel complaints." David Douglas added, "They assuredly produce flatulence: when in the Indian hut I was almost blown out by strength of wind."

When settlers began flooding into the Puget Sound region in the mid-1800s, the treeless prairies and the possibility of productive soils lured farmers and homesteaders. They soon converted thousands of acres to agricultural use for either crop production or as grazing lands for livestock. Pasture grasses took over native plant communities. Urban and residential development replaced the prairies, too. The final blow was the exclusion of fire. The elimination of fire allowed the widespread incursion of Douglas fir and the conversion of prairies to coniferous forests. Other invasives included grasses, nonnative fruit trees, and shrubs. Deep thatch and mosses built up in the absence of fire. This affected the germination and growing conditions for native plants and further changed the plant community structure.

Within a hundred years, the Puget Sound prairies had been decimated. Once covering an area three and a half times the size of Seattle, the remaining land covers less than three percent of its original expanse, seven square miles of the original 280 that lined the inland valleys from the Columbia River northward. The connectivity between discrete patches that once allowed birds, butterflies, seeds, and even fire to flow freely across the lowlands was lost. Much of the remaining habitat is of poor quality, triggering dramatic declines for native flora and fauna. Forty-six of the region's prairie plant species are on state or federal endangered/threatened watch lists. Nearly one-half of the 49 species of birds have undergone a population loss, a range reduction, or have been extirpated—become locally extinct.

Of the remaining 14,000 acres of Puget Sound prairies, about 11,000 lie in the lower Nisqually watershed on Joint Base Lewis-McChord (JBLM) between Tacoma and Olympia, at the center of some promising restoration projects and ground-zero for an intriguing irony. A prairie-oak ecosystem, among our rarest habitat, sits squarely on a heavily used military training site. Paratroopers, live artillery bombardments, and other explosions regularly rain down on the base's Artillery Impact Area, known historically as the Nisqually Plains. Nerve-shredding booms echo for miles like the soundtrack for a post-apocalyptic movie scene. It is hard to believe that by littering the ground with unexploded ordnance that includes depleted uranium shells, the war games actually contribute to high quality prairie. While counterintuitive, shelling the land with live artillery is a better option than converting it to agricultural, commercial, or residential use. Moreover, since fire is vital to prairie health, the high frequency, low-intensity fires caused by exploding artillery favor native plants while hindering the invasive, non-fire-resistant species.

The Army prefers high quality natural habitat for its training exercises, and prairies over forests, so it invests heavily to continue training on the JBLM sites. The U.S. Army and the Office of the Secretary of Defense committed over $330 million to the Army Compatible Use Buffer (ACUB) Program to ensure the continuation of training operations at 30 military installations around the country. The Joint Base Lewis-McChord ACUB Program and its partner The Nature Conservancy spent nearly $10 million to preserve the remaining habitat and its at-risk species. Together the two unlikely allies protected over a thousand acres on and off base at JBLM.

The Department of Defense (DOD) found a new partner in the Center for Natural Lands Management in 2011, which coordinates a wide variety of partners to conserve prairies throughout the ecoregion. In 2013, the DOD and the U.S. Departments of Agriculture and the Interior established the Sentinel Landscapes Partnership to promote nature resource sustainability and agricultural and conservation land use adjacent to military installations. The program designated JBLM as its pilot Sentinel Landscape. With collaborators that now include federal, state, and county agencies, land trusts and other non-governmental agencies, they have developed a conservation consortium that has brought attention, resources, and success to a collection of restoration projects on the south Puget Sound prairies.

THREATENED SPECIES, DRASTIC MEASURES

Full restoration of the Nisqually Valley and other south Puget Sound prairies is a long shot, but a dedicated and diverse group of partners keeps pushing to arrest and reverse some of the damage. Like most ecological rehabilitation projects, the work is a gradual, time-consuming process aimed at returning the ecosystem to its original condition, in this case, prior to the arrival of European American settlers. Species restoration seeks to return individual species to their earlier levels. Some of the most creative and painstaking efforts benefits those protected under the U.S. Endangered Species Act (ESA), which seeks to conserve threatened and endangered species and their ecosystems. Those listed as threatened have experienced such serious population declines and range contractions that resource managers must take action to prevent them from reaching endangered status. Species on the endangered list are in jeopardy of extinction across all or part of their range.

THE STREAKED HORNED LARK

A recent addition to the Endangered Species Act, North America's only established member of the lark family, has one of the most uncertain futures of the region's prairie birds. The streaked horned lark is one of 24 currently recognized subspecies of the horned lark. It lives in grasslands and is intermediate in size between a house sparrow and an American robin. Once flourishing from southern British Columbia to southern Oregon, its range has shrunk by more than half in the last hundred years. In addition to lost habitat, the exclusion of fire and increased flood control resulting from dam construction degraded other areas. Predation by American crows, northern harriers, western meadowlarks, and coyotes, among other species, also contributed to the bird's decline. By the mid-1990s, the largest of the few surviving populations in Washington lived on the JBLM prairies. Fewer than 150 breeding pairs occur at 17 locations in the state, four of which are on base. And while just 1,170 to 1,610 individuals hang on, data suggests that populations at most sites are stable or increasing.

Another reason for the bird's century-long downtrend stems from a lack of genetic diversity, which can increase the susceptibility to disease and lower reproductive success. An egg hatch rate of 44 percent and a fledging rate of 28 percent signaled inbreeding—and alarmed researchers. They responded by designing a project to increase genetic diversity. The partners include the U.S. Department of Defense, the U.S. Fish and Wildlife Service (USFWS), Oregon Department of Fish and Wildlife, Oregon State University, and the Center for Natural Lands Management (CNLM).

The project involves replacing eggs in JBLM nests with those of a genetically different population in Oregon's Willamette Valley. The idea is to increase diversity by bringing in genetically different eggs, tricking the birds into incubating those eggs and raising young biologically unrelated to them. The project has produced some nestlings that fledged successfully, including one male from a translocated egg that returned to breed at JBLM for four consecutive years. Even this single returning bird can help rescue the local lark population.

One April morning, I accompany two CNLM biologists as they survey for juvenile lark survivorship on JBLM's 13th Division prairie. As we walk some of the highest quality prairie in the Pacific Northwest, it is easy to imagine this day two hundred years ago. Rolling out like an emerald carpet toward Mount Rainier, this stretch hints at the floral explosion that will come in a few weeks. The mountain seems to tower just beyond our reach with a still-fresh mantle of late winter snow. We flush a nesting killdeer that feigns a broken wing, hoping to lure us away from its four cryptic eggs resting on a gravel bed. The vibrant song of the western meadowlark rings melodically. Suddenly, the drone of Air Force cargo planes and Chinook helicopters jar me back to the present.

The morning's highlight comes when several birds perform an aerial display called skylarking. The male sings as it circles high overhead, alternately flapping, gliding, and rising. Then, without warning, it descends in a steep dive. Skylarking helps it survey its territory and attract mates after returning in mid-February from wintering grounds on the Washington Coast, lower Columbia River islands, and other locations further south. Once paired, the female builds a nest, a shallow depression in the ground lined with soft vegetation. She chooses the north side of a clump of Roemer's fescue that will shelter the young from the broiling summer sun. Some also build "patios" of small pebbles and bits of dirt on the nest's north side, probably to help with thermal regulation and to keep vegetation low, aiding in predator detection.

Besides the genetic rescue project, other factors seem to be helping the lark's recovery. The 2013 JBLM Endangered Species Management Plan for the Streaked Horned Lark aims to boost population recovery both on and off base. Recent mild winters and springs contribute to productive years. So too do new conservation actions such as altering mowing regimes and avoiding nesting areas. The composite result is an improvement in nest survival, juvenile survivorship, hatching rates, and a glimmer of optimism for the streaked horned lark's chances of survival.

TAYLOR'S CHECKERSPOT BUTTERFLY

The loss of habitat that crippled streaked horned lark populations in the Nisqually watershed has had a similar effect on another winged species. The endangered Taylor's checkerspot butterfly also battles for survival. With a wingspan of just over two inches and a striking stained-glass pattern of black, reddish-orange, and cream, it lives a low-profile existence for much of its seven-month life cycle. Adult females emerge and become active in May, laying their initial clutch of up to 100 or more eggs on plants that include the native harsh paintbrush and the nonnative English plantain. The emerging caterpillars feed on the host plants until early summer when they enter diapause. This state of estivation—dormancy induced by heat and dryness—allows them to escape summer's heat by moving into the prairie duff. As the weather turns colder, they transition to a hibernative phase without resuming feeding. They remain in diapause until mid-January when they begin feeding again, building up a storehouse of chemicals that help them avoid predation. The caterpillars either complete their metamorphosis in the spring or return to diapause for another year, waiting for better weather conditions.

Emerging adults usually begin appearing in mid-April, but the timing can vary by several weeks. The entire flight period typically lasts only five weeks. Each individual's flight life spans no more than 14 days; about half live only a few days.

Like the streaked horned lark, Taylor's checkerspots were once plentiful throughout the Pacific Northwest prairies. But the number of sites dwindled to 15 by 1996, and they now occur at fewer than a dozen locations. Hoping to stave off the decline, the Washington Department of Fish and Wildlife collaborated with staff at Portland's Oregon Zoo to pioneer captive breeding techniques. They were soon rearing adults and attempting captive mating and egg production.

The project received a boost when the JBLM ACUB program and the U.S. Fish and Wildlife Service funded a breeding and rearing facility at the Mission Creek Corrections Center for Women at Belfair, Washington. With the assistance of outside experts and corrections staff, incarcerated women have raised over 18,000 caterpillars and butterflies for release since 2011 as part of the Sustainability in Prisons Project (SPP), a joint venture between the Washington State Department of Corrections and The Evergreen State College. With a variety of conservation programs in all 12 Washington correctional facilities, its mission is to "empower sustainable change by bringing nature, science, and environmental education into prisons." On a still-foggy June morning, I visited the greenhouse-turned-butterfly nursery to learn more. For these solar-powered organisms, the trick is to provide the perfect mix of warmth and sunlight, and the greenhouse's UV-transmitting glass is just the thing. One technician uses a mortar and pestle to grind plantain leaves into a "pesto" that she'll feed to the larvae.

Another uses a paper clip to hand feed a honey solution to adult female checkerspots. Some butterflies have already laid their yellowish eggs, the size of fine-ground pepper. I can easily see the value of the work on the butterfly's behalf, but I have to listen carefully to learn of its impact on the women. Feeding the adult females now in the final phase of their lives, one tells me, "They teach me that the end of life can be just as beautiful as other parts of life." Another mentions the facility's quiet calm and that, "It makes me feel like a human being. Working here gives me hope. We're saving their lives, and they've changed mine," she says. "It makes me think that the best is yet to come—for the butterflies, and for us."

The early success of producing eggs, rearing larvae, and raising mating adults, combined with habitat management, may help the checkerspot beat the odds on extinction. Its numbers have generally improved in recent years. Project staff may also reach a short-term goal of establishing populations at three different Puget lowland sites by 2022. In the meantime, Mission Creek residents find value in their work and meaning in their lives from their relationship with a brilliant butterfly that goes largely unnoticed for most of its life.

The Big Picture for Prairie Restoration

Hopes for the streaked horned lark and Taylor's checkerspot butterfly hinge on the success of efforts targeting those species, but these efforts represent only a sample of the work underway in the watershed. Besides species-specific activities, another approach aims to increase the overall health of prairie ecosystems. Researchers work to improve or restore ecosystem functionality by creating optimum conditions for long-term prairie health. One way is to remove invasive plants, none of which is more tenacious than Scot's or Scotch broom. Its removal is vital to prairie-wide restoration and requires a multi-dimensional, persistent attack. Introduced in the eastern United States in the early 1800s and on Vancouver Island in 1850, a single shrub produces thousands of long-lived seeds. Control efforts to reduce "old growth forests" of Scotch broom include mechanical and manual removal, the use of herbicides, and prescribed fire for up to 10 years. It's almost impossible to eradicate it completely. Other prairie invasives yield to spring and fall mowing and the use of grass-specific herbicides. The degree of success varies. Combining the removal of invasives with native seed collection and plant cultivation, however, holds some measure of promise for prairie ecosystems.

JBLM staff had already operated a greenhouse program, but production shifted into high gear when they began working with the Sustainability in Prisons Project to propagate native plants. It started at the Stafford Creek Corrections Center in Aberdeen, soon joined by an off-site crew from the Cedar Creek Corrections Center. Growth continued with the addition of the Washington Corrections Center for Women in Gig Harbor, and finally with the startup of a native seed nursery at the Washington Corrections Center in Shelton. Together, the combined facilities have propagated more than 2.5 million plants and produced over 25 pounds of native seed. Technicians developed and followed protocols that enabled them to grow, in nurseries for the first time, over 85 species of native plants. About half go directly into the prairies, clustered mostly as de facto butterfly gardens. Many serve as host plants for checkerspot larvae and as host nectar plants for adults.

The remaining plants become part of a seed production operation. Technicians grow nursery plants to maturity and then collect the seed by hand. Once the seed is cleaned and prepared, workers sow it directly onto the prairie. The survival rate for direct sown seed is lower than for direct transplants, but it enables workers to cover more ground than the labor-intensive transplanting.

In addition to the removal of invasive plants and the nursery-propagated native plants and seed, the return of fire to prairie landscapes increases the odds of large-scale restoration. I came of age in an era where Smokey Bear exhorted Americans that "Only YOU can prevent forest fires," and Smokey did his job well. Despite knowing that its use by Native Americans helped maintain prairies and that used wisely it plays a valuable role in restoration, I am fear-stricken by the power and noise of a 60-acre crackling inferno at Glacial Heritage Preserve. The ignition crew uses cans of diesel mix to set precise locations ablaze. Just when the fires seem about to roar out of control, the holding crew moves in to guide and redirect it. Smoky billows of gray and black spiral skyward as purple martins and barn swallows swoop and glide to feed on insects fleeing the blaze. It looks like chaos, but highly trained fire bosses choreograph every move to keep the sea of flames in check. They reduce unneeded fuels and kill the dreaded Scotch broom while mimicking nature as much as possible. The burn is an ancient ecological process that the crew replicates with fire and then follows with "rain," using water to extinguish the flames rather than fire suppression tools, such as Pulaskis, that scar the ground.

I returned the next spring to see firsthand the effects of prescribed fire on prairie lands. Camas and scores of other blooming plants speckled the lush prairie that bore no resemblance to the charred landscape of nine months prior. Taylor's checkerspots, on the rebound in this small swath, zigzagged erratically. Western meadowlarks sang sweetly from hidden spots. Savannah sparrows called from practically every exposed perch. It was a dynamic ecosystem full of vitality. I had learned to see fire not as Smokey would have me see it, but as a restoration ecologist might see it: as a sign forecasting life and biodiversity.

In the last decade, more than 650 prescribed fires—most at JBLM and averaging over 1,800 acres per year—have helped restore ecological function to south Puget Sound prairies. The burn season runs from August to mid-October; a three to five-year burn cycle appears to be ideal. Optimism runs high for fire's potential to re-establish high quality land. If current burn programs keep pace with recent years, one projection estimates that local prairies could increase their areas by up to one-third.

Though prescribed burns hold promise, the practice comes with risk and controversy. Fire intensity and vegetation structure and composition can have highly variable effects on prairie ecosystems. Some invasive plants also benefit from fire, which stimulates the seed banks of Scotch broom and hairy cat's ear, for example. Many climate experts predict warmer and drier summers accompanied by wetter fall weather, which could shorten the burn season.

Aside from the ecological uncertainties of prescribed fire, the effects on people can be wide-ranging and may limit its use. Prescribed burns will come under greater scrutiny as human populations and developed areas press up against prairie lands where fire is used. Burn bans, fire hazard ratings, air quality concerns, and research that links smoke pollution to asthma may limit or prohibit prescribed burns in the coming years. The use of fire may stop again, only this time for reasons different from those in the nineteenth century.

The modest progress made at the species and ecosystem levels on the south Sound prairies in general and in the Nisqually River watershed in particular is offset by continuing threats. Estimated population growth by over one-fourth between 2010 and 2030 will create even more pressure to develop open spaces. Without prescribed fire, Douglas fir and other woody vegetation will continue encroaching upon prairie lands. Newly arriving organisms and pathogens will likely join the existing invasive plants and animals and further modify the diversity and structure of prairie-oak woodlands.

Other challenges loom with respect to climate change. The increasing temperatures and rainfall predicted for the twenty-first century should not disadvantage prairies, however. They may lead, in fact, to expansion and an increase in wetland sites that could host rare plant species. One potential effect of climate change that troubles biologists is phenological mismatches similar to those anticipated by researchers in the subalpine meadows at Paradise. Butterfly species that emerge before their primary host plants bloom could suffer wholesale die-offs. Mismatches have been linked to butterfly extinctions elsewhere and further research should shed more light on the possibility of these events among local species. Whatever the future holds for the prairies, many experts believe that climate change may produce major losses for some native species.

Anyone involved in the arduous work of the scope and size of these projects needs a story to hold onto, a glimmer of promise. One comes from the western bluebird, once relatively common in western Washington. Like other species, it suffered from habitat degradation and loss. By 1981, just four pairs of birds nested. Western bluebirds nest in cavities, and they responded well to a nest box program. Within a decade, over 200 nesting pairs fledged more than 700 young. Up to 300 bluebird pairs now nest at JBLM and on other nearby prairies. The program was so successful that the birds became the source population for reintroduction programs on San Jan Island and in Cowichan Valley, British Columbia.

In summer's heat when the work grinds on, hope rides on blue wings. The western bluebird, the prairie precedent for ecological success via human intervention, offers promise for similar victories in the future.

RESTORING ANCIENT CONNECTIONS

The river begins brash and bold at the Nisqually Glacier, but empties quietly into Puget Sound. Languid here, it flows without a fight. The delta flushes and fills twice daily as it pulses to tidal rhythms. The Nisqually's fresh water mixes with Puget Sound's saltwater to produce an estuary, one of nature's most productive ecosystems. The last of Puget Sound's undeveloped major estuaries, the Nisqually estuary is the best example of salt marsh habitat in this corner of the world. The unique mix of nutrients and sediments makes a perfect petri dish for millions of organisms. Over 200 species of birds and dozens of mammals, reptiles, amphibians, and fish live on the estuary. The mudflats teem with invertebrates, which act as the engine that drives the ecosystem. Thousands of worms and freshwater shrimp live in every square yard of muck and mud.

I watch as a great blue heron stands motionless, a solitary hunter stalking forage fish. With scarcely a twitch, it strikes and wrestles a silvery morsel. The noiseless wader and the tranquil delta belie the steady hum of interstate traffic just a mile away. The rumble of semi-trailer trucks anchors me in the present-day, but the land tells its stories of long ago.

Nisqually villages once stood on both sides of the river mouth. People called the village on the south shore *dxʷsqʷali* for "place of hay." Another village stood at nearby Medicine Creek, known today as McAllister Creek after the family that settled there in 1845. The Indians called a spot on its south shore *sxʷdadəb*, Lushootseed for a place where people acquired a form of spirit power or "medicine." The villagers harvested and ate, preserved, and traded the marine bounty that the waters gave them. They ate more clams, geoducks, mussels, and oysters than did their upriver relatives. A common saying amongst Coast Salish saltwater people was that, "When the tide is out, the table is set." Fish ranged from the diminutive herring and smelt to the mid-sized cutthroat trout, rainbow trout, and salmon up to the mammoth flatfish, lingcod, and sturgeon.

In a fir grove near here in December 1854, Indian signatories from Nisqually, Squaxin, Puyallup, and other bands affixed their marks to the Treaty of Medicine Creek. They ceded 4,000 square miles of land to the federal government while reserving their right to fish, hunt, and gather. The longest living survivor of the proceedings, a mighty Douglas fir known as the Treaty Tree, endured another 150 years before falling in a windstorm.

Little changed at the river's mouth after treaty times. Then, in the early 1900s, its ecological processes changed drastically after Seattle lawyer Alson Brown purchased 3,250 acres. Brown's vision for a bustling farm began with a crew of 30 men and a horse-drawn scoop that dredged and built a four-mile dike intended to drain the delta. The dike created Brown's farmland but severed the river's natural connection to Puget Sound. Tides no longer rose and fell on the salt marsh and mudflats. The salmon lost access to off-channel habitat. Brown's enterprise eventually failed, but the damage to the estuary would linger for another hundred years.

Elsewhere in Puget Sound, other estuaries underwent conversion to commercial or private use as ports, marinas, or other facilities. By the 1950s,

Aerial photos showing the dike at Billy Frank Jr. Nisqually National Wildlife Refuge before and after breaching (top and bottom, respectively). Tidal action removed considerably more dike material after the lower photo was taken. The twin barns are circled to provide perspective. *Upper photo: Curtis D. Tanner/U.S. Fish and Wildlife Service. Lower photo: Brian Root/U.S. Fish and Wildlife Service*

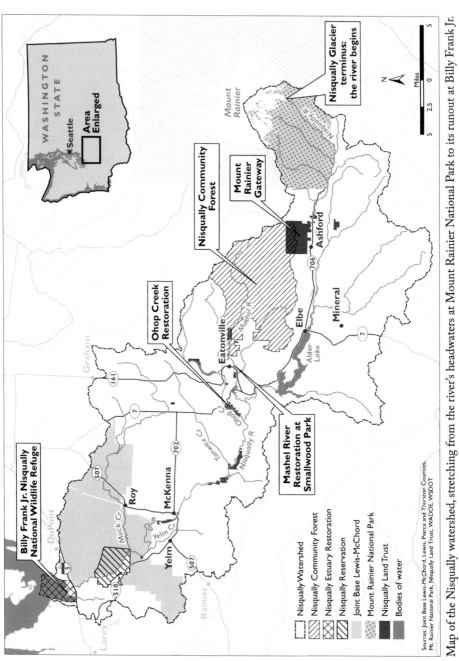

Map of the Nisqually watershed, stretching from the river's headwaters at Mount Rainier National Park to its runout at Billy Frank Jr. Nisqually National Wildlife Refuge. *Jennifer Cutler/GIS Program Manager, Nisqually Indian Tribe*

Sources: Joint Base Lewis-McChord, Lewis, Pierce and Thurston Counties, Mt. Rainier National Park, Nisqually Land Trust, WADOE, WSDOT

only 15 percent of the original Puget Sound estuarine habitat remained. The marine-rich waterways were under siege. In the south Sound, though, Olympia-area conservationists mobilized to head off the runaway development of shoreline areas. Their efforts led the charge for shoreline planning and permitting through the state's Shoreline Management Act of 1971. More work resulted in the establishment of the Nisqually National Wildlife Refuge in 1974, assuring protection for the delta proper in perpetuity. It made the Nisqually one of the rare rivers with its headwaters in a national park and its runout in a national wildlife refuge.

Day-to-day operations on the refuge focused on habitat management for resident species and those using it as a migratory stopover. Then, in the early 2000s, the theme of action pioneered by earlier conservationists enjoyed a resurgence. A group of partners that included the Nisqually Indian Tribe, the U.S. Fish and Wildlife Service, the U.S. Geological Survey, Ducks Unlimited, and other groups set their sights on a bold goal: remove the Brown's Farm dike and allow the Nisqually to reclaim its ancestral channels.

Years of planning, fundraising, and permitting finally paid off in the spring of 2009. Crews used excavators, bulldozers, and dump trucks to remove nearly five miles of dike. The work returned the estuary to its original grade. A new setback levee created 250 acres of freshwater habitat for waterfowl. The project increased the size of the south Sound's salt marsh habitat by 50 percent, restored 762 acres of estuary, and reconnected the delta to its original pathways. Other work brought the total to over 900 restored acres with nearly 20 miles of reopened tidal channels that showed a nearly identical pattern of estuarine connectivity with maps from 1878. Workers and volunteers installed 20,000 plants that included the salt-tolerant twinberry, clustered swamp rose, and pickleweed that would catch the occasional high tide.

Daily tidal action now sends nutrients and sediments swirling through the delta. Migrating shorebirds have a new refueling spot. Waterfowl use the freshwater pond. High tides swell and wash through a forested wetland, a refuge for salmon undergoing physiological changes between freshwater and saltwater environments.

The recovery of ecological processes on the Nisqually delta brings new questions and new challenges. Some researchers want to know more about its biological components like the vegetation and the invertebrate, salmon, and bird populations and their changes over time. Others study the physical elements that include water quality, erosion, and sedimentation at the estuary and in near shore environments. Their findings will help guide subsequent projects.

The looming question for the Nisqually Tribe rests on the future of the threatened Chinook salmon. Long-prized as "The King," Chinook make good use of the newly restored habitat. The tribe hopes that the reopened channels will increase its numbers in the watershed. If they succeed, it will revitalize another ancient connection.

FLOWING FORWARD

More than 30 years after Qu-lash-qud's son Billy Frank Jr. challenged the Nisqually watershed stakeholders to set aside their differences and work together, the Nisqually River Council continues its collaborative stewardship that shines as a national model. The oldest active river council this side of the Mississippi, its broad membership that includes the Nisqually Tribe, municipal and county governments, state and federal agencies, various advisory committees, and other allies moves forward together on a host of initiatives. This group of partners does its work with such good effect that the U.S. Department of the Interior honored the Nisqually River Council as a "blueprint for cooperative conservation projects."

Even with the high praise and a constellation of successes, the council carries on its steady pace of advocacy and action. Its Watershed Stewardship Plan sketches a 15-year program of specific action items. Researchers continue working to understand the short and long-term effects of climate change on the Nisqually River system. Others chart progress on restoring the five salmon species that run in the river, but more work—and more habitat—is needed. Moreover, since the fate of returning salmon hinges on conditions in Puget Sound and the open ocean, research must extend into Puget Sound and beyond. In addition to the Chinook salmon, Puget Sound steelhead, streaked horned lark, and Taylor's checkerspot butterfly, other federally protected species living in the Nisqually watershed include bull trout, the marbled murrelet, the northern spotted owl, the Mazama pocket gopher, and golden paintbrush. Nearly two dozen other plants and animals receive protection at the state level. All of these have the close attention of the Nisqually River Council and its members.

Beyond the 15-year plan, the council's vision extends into the 2050s. Seeking to "develop a place where people can earn a living, be part of a community, and enhance the environment," the plan aims to protect "the entire watershed—its people, its businesses, its economy, its tourism, its wildlife habitat, and its water resources."

On a perfectly gorgeous August morning, my wife Valerie and I have come to float a section of the Nisqually River. We are not rafters, but are encouraged by our professional guide's easy manner and buoyant confidence. We'll spend the next four or five hours floating leisurely down to our takeout point. The river's rapids pose little challenge for experienced boaters, but its pristine character and tranquil beauty attracts people of all skill levels.

We set off amidst towering black cottonwoods and big leaf maples that line the river. Red alder and vine maple fill in the lower gaps. Together these species make up the dominant woody plants of this riparian community. Within minutes, western tiger swallowtail butterflies join us. The gentlest of all tigers, they wing tirelessly off the bow as if to guide us downriver.

Spotted sandpipers busily ferry insects up and down the river's edge, food for their nestlings. Their incessant tail bobbing identifies them at a distance. As one of the few birds with reversed sex roles, the female takes the lead in claiming territory and in courtship. She often mates with up to four males, each of which then incubates the eggs. Males typically assume the bulk of the feeding, tending, and rearing responsibilities. The high volume of activity indicates that it has been a successful breeding season. Within a few weeks, the birds will head for the southernmost United States. Some will fly on to winter in South America. Most sandpipers migrate in large flocks, but spotted sandpipers migrate singly or in small groups.

Patrolling broods of common mergansers, one of the river's most common ducks, use the overhanging riverside foliage to stay out of sight. A mated pair of belted kingfishers rattles wildly, strengthening their pair bond. Cedar waxwings, delicate, feathered acrobats, prey on insects mid-air and mid-river. With scarcely a sign of human habitation on this stretch of river, I reflect on the watershed's communities, plants, animals, projects, and people. The person who best personified the indomitable spirit of the Nisqually was tribal leader and activist Billy Frank Jr., worked to benefit the salmon, the river, and all its inhabitants. His influence and example earned him the Presidential Medal of Freedom, bestowed posthumously by President Barack Obama in 2015. Fittingly, in 2016 the refuge was renamed the Billy Frank Jr. Nisqually National Wildlife Refuge. A powerhouse of a man with a big hug and a warm smile, he bound together the aspirations and dreams for those living and working in the watershed when he said,

I don't believe in magic. I believe in the sun and the stars, the water, the tides, the floods, the owls, the hawks flying, the river running, the wind talking. They're measurements. They tell us how healthy things are. How healthy we are. Because we and they are the same. That's what I believe in. Those who listen to the world that sustains them can hear the message brought forth by the salmon.

Billy Frank Jr. with the Treaty Tree, shortly after it blew down in a December 2006 windstorm. *Ellen M. Banner/Seattle Times*

4

Historic Longmire and Surrounding Area

Mount Rainier from Longmire, near the Community Building. Red alder, foreground, has colonized the area since the 2006 flood. *Drawing by Lucia Harrison*

LAND OF DISTURBANCE

The sunny promise of a May morning filtered through the translucent walls of our two-person tent. Outside, the Douglas firs that crowded the Longmire Campground stood against a blue-sky backdrop. We had slept in and the sun had already crested the backbone of Eagle Peak. An extraordinary day lay ahead.

Valerie and I had just returned to the Pacific Northwest from a cross-country trip, and were excited to be home. As we chatted about our plans, a low rumble, like a distant log truck, intruded on our conversation.

"Probably just someone driving across the bridge," I mused. We made ready for the day and noticed that the bright morning had suddenly turned over-cast. Considering that the mountain is prone to abrupt weather changes, this was no surprise.

We dressed for breakfast as raindrops began dancing on our tent. "That was quick," I said. "That's the fastest it's ever changed from sunny to rainy on us." Within moments, we heard voices calling loudly and car doors slamming. Suddenly a car made a quick stop just outside our tent. I poked my head out to find a park ranger standing a few feet away. "You folks will have to pack up and leave the park as quickly as you can," he blurted. "Mount St. Helens just blew her top! Mount Rainier is closed!"

We scrambled for our gear, threw it helter-skelter into the car, and began driving toward the Nisqually entrance. Within minutes, morning became night as a dense, expanding ash cloud hid the morning sun. What I thought were cars rattling the timbers of the bridge was Mount St. Helens's rumbling, explosive blast. The drops of "rain" on our tent were fragments from the neighboring Cascade volcano, blasted 15 miles into the stratosphere, drifting earthward. Our car crawled forward, headlights peering into the eerie gloom. The wipers labored to clear the falling ash off the windshield. The near-zero visibility and complete darkness at mid-morning sent a single, somber message: *This is what it is like when the world ends.*

We were among the last people to leave Longmire, and the deserted road heightened the apocalyptic feel of the premature darkness. After what seemed an eternity to reach the park entrance, we arrived at a cabin where friends awaited us. We stood amid the falling ash in a scene as silent as a January midnight snowfall. An inch or more of the fine, gray grit had already accumulated. Not since south central Oregon's Mount Mazama had exploded some 7,700 years earlier to create Crater Lake had this area been so affected by another mountain's volcanic eruption.

Mount St. Helens' cataclysmic eruption on May 18, 1980, captured the country's full attention. Besides learning about Harry Truman, the color-ful, recalcitrant lodge owner who refused to evacuate, people became aware of the geologists engaged in the rare opportunity to observe a brewing, bulging mountain. Lessons learned from the 57 lives lost in that eruption have since saved many others. In 1991 at Mount Pinatubo in the Philippines, the use of real time data and careful monitoring resulted in an accurate prediction of the eruption and the lahars that followed. Those living within the hazard zone evacuated in time, saving thousands of lives.

In the wake of Mount St. Helens, geologists can better read the deposits on Mount Rainier's flanks that are crucial to understanding its prior eruptions. They also have a better idea about the causes of debris flows and can more easily calculate their likelihood, type, and potential size. With this information, geologists can recognize warning signals and possibly tip-off local authorities quickly enough to enact evacuation procedures.

The Mount St. Helens blast was a reminder of the scope and scale of natural disasters that occur in our region. In addition to volcanic eruptions and their accompanying debris flows, they include earthquakes, fires, floods, tsunamis, and windstorms. Called disturbance events, they result in a sudden loss of biomass or changes in ecosystem structure or function, and play a central role in shaping the Longmire landscape. With the skill of crime scene investigators, geologists have pieced together clues to chronicle Longmire's disturbance events over the ages.

As part of his field research in the 1950s and 1960s, pioneering Mount Rainier geologist Rocky Crandell gathered evidence to describe the Paradise lahar that rumbled down the Paradise River equivalent to the height of an 80-story building. It emptied into the Nisqually River and eventually covered Longmire with about four feet of debris. Other geologists later amplified Crandell's work to age the lahar flow at around 5,600 years, leading to speculation that it may have been part of the same series of volcanic events as the Osceola Mudflow. Crandell also uncovered ash about an inch thick from a Mount St. Helens eruption that later studies dated to 1479. He found proof that Mount Rainier lahars covered the area after 1480. Later sleuthing of tree ages narrowed the date of debris flow deposits to between 1479 and 1570. More careful research established that another debris flow rolled through the Longmire area between 1686 and 1700, this one likely responsible for the uneven landscape near the Longmire Community Building and Longmire Stewardship Campground. From other signs that included boulder levees and relatively young stands of trees, Crandell concluded that yet another lahar churned through at around 1860. Disturbance events like these are the prevailing story at Longmire, and nearly every building—including those that make the area a National Historic District—is built upon lahar deposits. Lahars have shaped the Longmire landscape, and others are sure to follow.

A park employee surveys the 2006 flood damage to the Nisqually-Paradise Road at Sunshine Point, just inside the Nisqually Entrance. Note the root wads in the river channel in the background. In the lower photo, floodwaters in the maintenance area at Longmire nearly swept structures downstream. *Courtesy of Mount Rainier National Park Archives*

Another force of nature regularly brings menacing danger and anxiety to park staff. Almost every winter, storm systems form far out over the Pacific Ocean that can produce enormous amounts of precipitation. These gigantic plumes of warm water vapor originate in tropical latitudes, growing to more than 1,200 miles long and up to nearly 600 miles wide. Known as an atmospheric river (AR) or "pineapple express," the phenomenon can produce twice as much rainfall as typical storms. Its warmth brings rain instead of snow to high elevations, creating huge volumes of run-off. These downpours are responsible for many of the region's floods, and one that dumped 18 inches of rain in 36 hours in early November 2006 was the most intense of 119 storms of its kind in a nine-year period. It came on the heels of other ARs that had already fully saturated the ground, leaving the soil unable to absorb any more water. Runoff emptied directly into Mount Rainier's streams and rivers to cause the largest flooding event of its kind in park history.

People driving to work at Longmire on November 7, 2006, found Tahoma Creek choked with debris. Further up, Kautz Creek had forged a new channel, leaving the park road partially submerged. Later that morning, several employees looked on as trees and other debris rafted down the Nisqually River. They watched the riverbank erode in front of the Longmire Community Building, the red alders and Douglas firs whirling drunkenly before crashing into the teeming river. The staffers then wisely decided it was time to head for safety. Everyone evacuated Longmire and the park successfully, but not before water lapped up against the wheel wells of the last car across Kautz Creek.

Damage to the park was extensive, forcing it to close for the first time since the 1980 Mount St. Helens eruption. Every major road sustained damage, including the Nisqually-Paradise Road. The flood swept away bridges, foot logs, and washed out about 12 miles of trail. Most of the Sunshine Point Campground and Picnic Area floated downstream. It destroyed protective levees along the Nisqually River in Longmire and nearly washed the park's Emergency Operations Center downriver. It transported an immense volume of sediment, the equivalent of over 16,000 dump trucks. Aggradation ranged between five inches and five feet of added material to some sections of river. Debris contaminated the park's water system. Severed power and sewer lines needed repair. With damage between $24 and $27 million, the park remained closed for nearly six months as staff and contractors scrambled to restore operations. Over a two-year period, almost 2,000 volunteers per year blistered, grunted, and perspired through nearly 160,000 hours of conservation service as part of the Mount Rainier Recovery Corps, a value of $3.1 million. The consortium of non-governmental organizations and corporate partners was so effective that it won a U.S. Department of the Interior Cooperative Conservation Award.

The heavily eroded banks of the Nisqually River show ample evidence of the disturbance-prone nature of the Longmire area. Red alder saplings and other early successional species colonize recently disturbed areas along the river. Skunk cabbage and horsetail accompany red alder in low, wet areas. Other telltale signs include old flood channels and naturally occurring levees. In the 130-year history of the built environment at Longmire, people have frequently dealt with high water and erosion, yet have continued to build on the floodplain. Any notion of the area as static and unchanging ignores the dynamic nature and power of lahars and floods to alter landscapes at a moment's notice.

Another byproduct of Longmire's disturbance events is the dearth of archaeological evidence. Regular deluges and debris flows covered or swept away nearly all signs of precontact activity. But based on the ancient, well-established trails and the fact that native people guided European Americans there, archaeologists believe that indigenous people frequented the area over the ages.

Longmire Gets its Name

Development at Mount Rainier began with James Longmire in the 1880s. According to the oft-told story, he stumbled onto the meadow's mineral springs while searching for his horse, following his climb of the mountain in 1883. Recognizing a business opportunity, Longmire filed an 18-acre claim and soon employed locals to build a wagon road from present-day Ashford to the springs. In 1885 he hosted his first few intrepid, cash-paying customers and by decade's end, bathhouses and cabins awaited guests at "Longmire's Medical Springs." The family's grip on the area's commerce tightened, and eventually they built the Longmire Springs Hotel and an addition. Developers leasing property from the family built a new hotel and sixteen cottages in 1916. Several years later and under new ownership, the hotel was renamed the National Park Annex. It became part of the National Park Inn, which burned down in 1926. Undamaged, the Annex was renamed the National

Park Inn. Today's inn is a remodeled version of the original Annex.

The Longmire family's relationship with the park's administration eventually soured, and they sold their interests to the federal government in 1939. Their imprint lingers on the Trail of the Shadows, just across the road from the National Park Inn. A plaque marks the location of the Longmire Springs Hotel, and a log cabin replica and two developed mineral springs remain intact. Soda Springs is notable for its gas content that is almost entirely carbon dioxide. Iron Mike's reddish color comes from the iron content of the water.

The Trail of the Shadows leads to some interesting natural history worth investigating. The mostly flat loop trail encircles Longmire Meadow, home of the mineral springs that induced Indiana native and Yelm Prairie homesteader James Longmire to develop the area. At one time nearly 50 springs bubbled in the meadow, but beaver activity and the ensuing floods reduced the number by more than half. The water temperature of the springs has held at a steady 82° Fahrenheit over the last hundred years.

Longmire Meadows and the Longmire Springs Hotel, circa 1904. The rock structures in the foreground far left and far right enclose mineral springs for bathers. *Courtesy of Mount Rainier National Park Archives*

Next to the trail in the western portion of the meadow, visitors can see mineral deposits created by calcium carbonate and other minerals. Known as tufa and travertine, these deposits form when carbon dioxide escapes into the air. Although less ornate than similar features at Yellowstone National Park, they form modest mounds in the meadow. Their deep brown color comes from iron-rich minerals. At the Longmire Museum, young park visitors can find instructional materials and activities that introduce them to the mineral springs and provide opportunities to measure and record the water temperature.

The museum, built in 1916, was the park's first administration building. It became the park museum when a new administration building opened across the road in 1928. The 1928 building, used today for ranger offices and the Wilderness Information Center, along with the gas station just west of the museum and the community building across the Nisqually River, are classic examples of "park service rustic" architecture, which sought to blend in with natural surroundings. These early building were central to the park's designation as a National Historic Landmark District in 1997.

Nature's Wetlands Engineer

I rise before dawn this July morning, eager to search for one of the Long-mire Meadow's most elusive inhabitants. The American beaver is most active at dawn and dusk, so this is my best chance to get a glimpse of the creature that has had a formidable impact on North America's history. As I approach the meadow, first light gives shape to the day. Mist rises from the water. Varied thrushes give their lunch whistle call while American robins endlessly repeat their "cheery-up, cheery-oh" refrain. I soon make out the silhouettes of Vaux's swifts, barn, and violet-green swallows hawking insects overhead. I may not be lucky enough to spot today's target species, but I easily find its telltale signs. A western red cedar lies at the water's edge, looking as if an axe had felled it. Several other trees show girdling near their bases. A carefully constructed dam arcs gracefully across one section of open water. This is probably the best place to see a beaver, but luck and timing are essential.

I settle into a spot from which I can observe the largest patch of open water and wait. Controlling my breathing and sitting motionless, my patience is soon rewarded—I catch sight of a beaver at close range. Perhaps aware of me, it slaps its tail, its primary alarm system, and swims in the opposite direction. It stops to inspect a section of dam and moves a few sticks. The water flow increases noticeably. Slipping into the water it swims away, leaving a gentle wake.

This is North America's largest rodent, second in the world only to South America's capybara. Most range between 40 and 50 pounds and are well adapted to life in water. An excellent diver, it uses oxygen efficiently to remain underwater for up to 15 minutes. Its nostrils and ears close to keep water out and inner eyelids called nictitating membranes work like diving goggles. Lips close behind massive incisors to allow it to gnaw wood underwater. Large, webbed hind feet aid in swimming, but the multipurpose tail is what makes it a one-of-a-kind freshwater mammal. Flat and scaly, the broad tail becomes a rudder in water. It also stores fat and functions like a heat exchanger with a network of blood vessels that help reduce heat loss. One Native American story tells of beaver trading for his tail from muskrat, who was chief of the swamps. In another tale, "Wolf Brothers Kill Elk and Beaver," the four wolf brothers hunted the Big Beaver with ten tails. With the help of Honeysuckle and Scouring Rush, they captured it and cut it into small pieces. Hummingbird and Bumblebee then put pieces of beaver meat into every creek and lake. Beaver grew there, and thus became plentiful.

Once abundant from coast to coast and from Alaska to Mexico, the beaver's fortunes have intertwined with its perceived economic value for centuries. People first traded it off the Newfoundland coast in the late sixteenth century; the colony founded at Plymouth in 1620 began as a fur trading post. England's Hudson Bay Company (HBC) began operations in 1670 and for the next 200 years controlled most of the North American fur trade in a vast territory that eventually encompassed three million square miles. Although Captain Cook's search for a Northwest Passage failed in 1779, the expedition made huge profits on beaver pelts purchased in Nootka Sound off Vancouver Island that later sold in China. Thus began the highly profitable business of beaver skin trade in the Pacific Northwest that was already well established in the east. At that time, its pelts were valued on three continents and reaped prodigious profits. When the beaver hat gained popularity in Europe, the demand climbed even higher. A principal aim of Lewis and Clark's Corps of Discovery, in fact, was to blaze a trail for fur traders to the Pacific Ocean. They kept precise records of all sightings and signs of beaver encountered during their expedition.

In addition to its lush, warm pelt, beaver were sought for the various medicinal properties ascribed to a secretion from its paired anal glands known as castoreum. With a somewhat sweet odor imparting a leathery scent, perfumeries processed castoreum to use as a base in some perfumes.

In the 1820s, the Hudson's Bay Company realized that migrating Americans were destined to settle west of the Mississippi River and that it would soon lose its grip on the fur trade there. Convinced that the land would be less attractive to settlers if there were no beavers, the company undertook a policy to create a "fur desert" and eradicate the animal from the region. They succeeded. The HBC harvested and sold 4.7 million pelts by 1868, decimating the beaver in North America. American hunters and trappers not employed by the company also contributed to its precipitous decline. By the turn of the century, the American beaver population had crashed throughout most of its original range. Pushed to the edge of extinction in the Pacific Northwest and elsewhere, game management and reintroduction programs began in the early 1900s. Game regulations, coupled with a depressed market for pelts, boosted its recovery. It soon proved to be as resilient as it is industrious and quickly re-colonized much of its former range. Although uncommon in Washington until the 1930s, the American beaver is now widespread in wetlands around the state. They number about 20 million animals across North America.

The beaver's ecological importance matches its historical significance wherever it builds its lodge and dams. Beavers build either freestanding lodges or ones that connect to a bank with a burrow. Some forego a lodge and instead tunnel directly into the bank. The standard family of six includes a permanently bonded pair, the young of the year, and two yearlings. They are non-migratory and remain active year-round. Due to their size and their habitat, they have few enemies—except humans. These strict herbivores typically favor the inner bark of willow, cottonwood, and red alder, although they appear to settle for western red cedar in the Longmire Meadow. Beavers stockpile poles and branches underwater as a food cache that tides them over during the winter months. These traits, combined with their natural industriousness, make the beaver a wetlands engineer of the highest order. Their work felling trees helps create and maintain wetlands that keep the water table high. Some types of vegetation flourish in wetlands while others drown. Trees remaining as standing dead wood make valuable nest sites for other organisms. This structural diversity creates habitat for insects, fish, birds, and other mammals. Beavers play such a beneficial role in wetlands health that Idaho state game officers once parachuted them into new territories where they quickly adapted to their new environs and soon began building dams. Water held behind beaver dams for extended periods allows the soil to absorb toxins and excess nutrients, improving water quality. Beaver dams hold so much water that the U.S. Forest Service considered using their reservoirs to help fight forest fires. The dams are also leaky, making them more effective than concrete dams because their gradual release of water helps control flooding. Moreover, the dams stabilize riverbanks, reduce stream velocity, and allow the accumulation of sediment and vital nutrients like nitrogen, potassium, calcium, and iron.

The influence of beavers upon wetland systems has led biologists to consider it a keystone species. Similar to a keystone in the crown of an archway upon which the other stones lean and depend, other organisms rely upon beavers. Remove a keystone from an arch, and it may crumble. Removal of beavers causes cascading changes in the ecosystem's structure and function.

I complete my field notes, pleased with my good fortune so early in the day. As I finish my morning's work and head around the Trail of the Shadows, I notice several cavities excavated from stumps and snags, the handiwork of another keystone species, the pileated woodpecker.

Nature's Woodland Excavator

While enjoying supper in the dining room at Longmire's National Park Inn, I had the pleasure of an unexpected visitor. Just outside the window a few feet from my table, a female pileated woodpecker lit briefly on the flank of a Douglas fir. Except for the probably extinct ivory-billed woodpecker of the southeastern swamplands, the pileated is North America's largest woodpecker. Crow-sized, and sporting an unmistakable red crest at all ages and in both sexes, it takes its name from the fiery feathers covering the top of its head from bill to nape, known as its pileum. Pileated woodpeckers are non-migratory, permanent residents from eastern Texas and Oklahoma east to the Atlantic seaboard, north to New England and Canada, then west to Idaho, Washington, Oregon, and California. They do not regularly occur on the Great Plains, in the Great Basin, or in the arid Southwest.

Well adapted to late successional deciduous, coniferous, and mixed forests, pileated woodpeckers also select younger forests that include large dead or decaying trees. Adults pair for life and defend their territory year-round against other pileateds by drumming, calling, or chasing. If one of a mated pair dies, the remaining bird pairs with another from an adjoining territory. Clutch size is usually four; young adult birds often take up territories not far from where they were born.

The big ecological story behind these woodpeckers is in the impact they have on forest ecosystems and on the other animals living there. Although they feed on fruits and nuts to a small degree, pileated woodpeckers rely mostly on carpenter ants, wood boring beetle larvae, and dampwood termites that they find in decaying, large-diameter snags. They locate their prey by sight or by sound, responding to the movement of carpenter ants hidden within a tree or log after drumming on it. Powerful excavators, they use their long, sturdy, chisel-like bills to create rectangular foraging excavations up to eight inches wide and over 12 inches long. They sustain a life of pounding and hammering with shock absorber-like features that include a thick-walled skull with strong muscles and a tough membrane surrounding the brain.

Pileateds create nesting and roosting cavities in large dead or decaying trees within a home range that can exceed two and a half square miles. They avoid trees without heart rot that might be too demanding to excavate, as well as those in advanced stages of decay that might fail to support a nest and its activities. A pair may prospect at a number of different spots each spring before completing a new nest cavity. Each bird also excavates roosting sites in hollow trees. They use up to seven sites in the nonbreeding season, each with multiple openings that allow them to evade predators like Cooper's hawk, northern goshawk, great horned owl, or American marten. The birds fly to another roost when flushed.

As the forest's principal cavity excavator, pileated woodpeckers provide nest and roost sites for at least 38 species. Those using pileated-created cavities include squirrels, bats, American marten, American kestrel, ducks, owls, other species of woodpeckers, and even wasps. Weaker woodland excavators like downy and hairy woodpeckers benefit from pileated foraging sites by feeding on invertebrates that they otherwise couldn't reach with their shorter bills. Pileateds provide other ecosystem services, too. A diet that includes wood-boring beetle larvae may help control beetle populations and prevent widespread damage to trees. They play a role in forest decomposition and nutrient cycling by breaking up rotting wood and exposing it to the elements, which aids in the decomposition of snags and logs. They expose live wood to insect and fungal attacks and may help spread heart-rot fungi, which speeds up the forest's recycling processes.

In the same manner that the American beaver affects wetlands beyond its size or numbers, the far-reaching ecological influence of the pileated woodpecker earns it the label of a keystone species. Some researchers regard pileateds as keystone habitat modifiers because their activities change the physical structure of ecosystems. Despite habitat loss and fragmentation over much of their range and varying estimates of their population, they are increasing in the central and western United States. The largest population increases appear to be in the northwest and northeast regions. Wherever they occur, these ecosystem engineers continue to alter and shape woodland systems at Mount Rainier and elsewhere as few other organisms do.

THE ANCIENT ONES

Pileated woodpeckers inhabit a wide variety of woodlots and woodlands, but they prefer deep, mature forests like those at Longmire and other lowland areas of the park. Despite its renown for year-round snow cover and being the most glaciated peak in the contiguous United States, forest covers 60 percent of Mount Rainier. Commonly known as the lowland forest or western hemlock zone, it is also called the *Tsuga heterophylla* zone after the western hemlock's scientific name. *Tsuga* comes from the Japanese "tsu" meaning tree and "ga," mother. Other major species in the western hemlock zone include western red cedar and Douglas fir. The fir is a pioneering species that quickly colonizes an area after disturbance events like debris flows, disease, fire, or floods. When there is a long interval between events, the shade-tolerant hemlock eventually replaces it. Devil's club and sword fern are the most conspicuous understory plants, while the edges of waterways host bigleaf maple, black cottonwood, red alder, vine maple, and various willows. The zone reaches to about 3,000 feet in elevation, making up the most extensive vegetative zone in Washington and Oregon. Nearly two-thirds of the stands in the park are old growth forests—200 years or older. Old growth is relatively easy to find along the road leading into Longmire from the Nisqually entrance, on the west side of the Trail of the Shadows, and partway up Longmire's slopes away from the occasional damage caused by floods and lahars.

There was a time when old growth enshrouded most of the Pacific Northwest. A maritime climate featuring wet and mild winters with annual precipitation between 60 and 120 inches interrupted by dry summers helped establish the fast-growing forests. From the water's edge to the glacier's reach, old growth forests blanketed the region to create the densest forests on the continent. Only Oregon's Willamette Valley and the Puget Sound prairies were free of forests when European American settlers began arriving in the 1800s.

Old growth trees range between 150 and 250 feet tall in perhaps the greatest accumulation of biomass in the world. A single tree can have over 60 million individual needles that weigh in excess of 400 pounds. Old growth was an imposing impediment to agriculture and other development, but it also represented a resource of tremendous economic value. Logging the forests quickly became standard practice and the stuff of legends, continuing into the twentieth century, and surging after the World War II housing boom. Over the next 50 years, the region's old growth forests declined in area by two-thirds. Today, wilderness areas, national parks, and other reserves are the best locations to enjoy these living marvels. They are without equal in terms of size and splendor among the world's softwood forests.

On a walk in one of Mount Rainier's old growth forests, I came upon a live Douglas fir, its diameter at around nine feet. Nearby lay two recently downed giants, one a fir and the other a western red cedar. Windthrow had probably brought them down. Because they had obstructed the path, the trail crew had cut them out. Examining the fresh cuts, the fir measured 44 inches in diameter. Carefully counting its rings, I found it to be approximately 350 years old when it toppled to the forest floor. The cedar, at 42 inches in diameter, held over 450 rings. Tree rings are the forest's history book, and when this tree made its start as a sapling, Coast Salish cultures thrived in an abundance economy built around cedar and salmon. The Pacific Northwest was still 200 years away from its first visit by European American explorers. My ancestors would not arrive in America for another 300 years. Reflecting on the tree rings and its timeline, I felt a tiny piece of a larger whole, somehow connected to the spiral of life.

Returning to the immense, live Douglas fir, I pondered this shrine of earth and sky. Here, old growth trees burrow their serpentine roots into the soil to spread laterally in search of nutrients and moisture. They extend their tapering branches skyward, ever reaching for the life-giving sunlight that powers photosynthesis. These stalwart old trees hold their ground for centuries. Legions of saplings, assuring a cycle of regeneration, vie for available sunlight filtering through the canopy that makes its way to the forest floor. Downed logs scatter about like jackstraws in various stages of decay.

The term "old growth" came into use well over a hundred years ago, but there is little agreement over a single definition. It is difficult to determine exactly when a forest becomes old growth, a concept complicated by a variety of traits that include species makeup, site characteristics, and other factors. Most ecologists agree that old growth is a forest in its later stages of successional development—a minimum of 175 to 200 years old—possessing a suite of structural characteristics not found in younger forests 75 to 100 years old.

I sit near the downed logs whose rings I had counted to inventory the structural traits that make up this old growth forest. Once described as biological deserts to help rationalize clear-cuts and single-species row plantings, research now shows them to be highly productive, valuable ecosystems. A behemoth western red cedar, whose circumference I measured at 25 feet, looms over me. Joining neighboring old growth Douglas firs, they are the most obvious and impressive structural characteristic of the forest: giant trees of multiple species. I spy a pair of ravens settling into the top of a standing dead tree some 75 or 80 feet tall. Known as a snag, it's another important component of old growth stands. It provides habitat for both invertebrates and vertebrates until it crashes earthward, joining the other logs scattered there. These downed dead trees, called coarse woody debris, comprise another characteristic of old growth. The number, mass, and volume of downed logs often make up a significant portion of the forest's biomass. They contain essential nutrients like phosphorus and nitrogen and host bacteria that make it available to other forest organisms.

The woody debris lying in the nearby river is yet another character trait of these places. It stabilizes streambeds and banks, controls sediment and water flow, and influences the health of the forest's aquatic ecosystems. Woody debris creates dams and plunge pools that ensure structural diversity and an abundance of habitats.

In this one small patch, I found all of the structural characteristics of Pacific Northwest old growth forests: mammoth live trees with multiple layers of canopy, the standing dead wood known as snags, and downed logs on the forest floor and in streams. These traits don't typically occur in younger or managed stands. Together they work to deliver a variety of ecosystem services that include shade for streamside habitat, water filtration, habitat for numerous organisms, and nutrient cycling. Old growth forests also remove large quantities of carbon dioxide from the atmosphere, storing it in both live tissue and decomposing organic material. Once believed to be stable and little changing "climax" forests, we now understand them to be dynamic systems that change frequently as a result of continued growth, old age, and stand-replacing disturbances that include fire, windthrow, and insect outbreaks.

The old growth forests that we see and admire evoke wonder and awe, but their many unseen factors are equally impressive. Giant logs on the forest floor host scores of organisms engaged in the work of returning the log to the soil. Some trees have fallen recently and just begun the unremitting process of decomposition. Others have lain upon the ground for 500 years or more, having nearly completed their return to the soil. This decay and return, decomposition and regeneration, relies upon the work of myriad organisms that participate in the breakdown of a fallen tree. In a single Douglas fir, nearly 300 species of insects bore, chew, and grind the wood. Each has specific predators, parasites, and scavengers associated with it. Joining these invertebrates, other organisms include bacteria, fungi, plants, and vertebrates, all interacting in multiple webs of producers, consumers, and decomposers. Some depend completely on downed trees for survival. For all of them, it is home.

The decomposition process begins as bark-boring and wood-boring beetles tunnel through the outer bark to consume the tree's inner bark and sapwood. Carpenter ants, wood-tunneling mites, and Pacific dampwood termites join the beetles to create galleries and tunnels. The termite's gut protozoa digest cellulose, allowing it to consume the otherwise indigestible wood. The tunnels fill with borings and feces, known as frass. These droppings contain bacteria that thrive in the damp wood and attack it through enzymatic activity. Fungal spores hitchhike on some insects to access the log's inner reaches. The spores germinate, grow, and establish fruiting bodies. Many remain inside the rotting log. Bacteria and yeasts establish themselves on the fallen tree; other microorganisms and fungi were already at work before it fell. Tiny insects known as springtails, their locomotion powered by an abdominal spring-like mechanism, feed on the tree's microorganisms. Ambrosia beetles join the springtails. Checkered beetles come to feed on the bark-boring beetles. At an advanced level of decay that may take several hundred years, plants establish themselves on the rotting log. These are the nurse logs, easily found in Pacific Northwest old growth forests. Western hemlock, huckleberry, salal, and other species grow upon them.

The log continues its decay; its food web continues to thrive. Mites and insects such as the common earwig feed on the plants now established on the fallen log. Slugs and snails feed there, too. Other organisms like predaceous mites, one of the smallest forest predators, have a wide-ranging diet. Long-legged and quick moving, their strong mouthparts enable them to

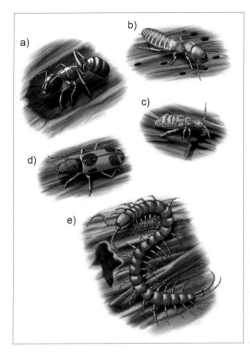

Some of the common insects found inside decomposing logs in the region's old growth forests: a) Carpenter ant; b) Pacific Coast dampwood termite; c) Slender springtail; d) Checkered beetle (no common name) (*Enoclerus eximius*); and e) Predaceous centipede (no common name) (*Scolopocryptops spinicaudus*). *Kirsten Wahlquist*

An adult northern spotted owl pounces on one of the park's survey crew-raised lab mice. *Emily Zazz Brouwer/National Park Service*

capture and chew springtails, small roundworms, or each other. They also eat plants, lichens, fungi, and bacteria. Pseudoscorpions lack the tail and stinger of true scorpions but still prey on a variety of insects. Predaceous centipedes with poisonous jaws paralyze their prey and play an important role in the later stages of tree decay.

Still more organisms, largely unseen, figure in the decay of an ancient log. Coastal giant salamanders, the largest in the Pacific Northwest and ranging to nearly a foot in length, inhabit nearby streams and downed logs in the advanced stages of decay. Small mammals like the Trowbridge shrew, the most common shrew in the forest, become active after dark. Resembling a mouse except for a long, pointed snout and oversized ears, its high metabolism requires it to eat the equivalent of its weight each day. It uses its surprising quickness to capture and eat dangerous prey like centipedes and wasps. The shrew's broad diet also includes spiders, slugs, snails, and at least several dozen other organisms.

Detritivores such as mites, millipedes, earwigs, and earthworms play a critical role, eating dead plant and animal materials. Finally, decomposition is complete, and the downed log has returned to the soil.

The Soul of the Forest

Longmire Village's utility area sits east and south of the gas station and administration building, bordering the Nisqually River. In addition to warehouses, workshops, and an automotive repair shop, it houses offices for the park's Natural and Cultural Resources staff. Sizing up the youthful, fit spotted owl survey crew this August morning, I am a bit intimidated. Late in the field season, they've spent the last month or two hiking long distances and carrying heavy equipment to research sites. We'll be moving quickly through the forest and off-trail over steep terrain, and I dread getting lost or slowing them up. Their friendly welcome reassures me.

Before we set out on foot for the lower flank of Eagle Peak, I get a briefing over the park map about the area we'll be surveying. On our way out the door, I visit the "mouse house," a closet-sized affair set off from an office, where the owl survey crew raises white lab mice. One of the crew catches several—today's sacrifice. She puts them in a container with air holes and chucks it into her daypack.

Given the cold, wet spring, it has not been a particularly good year for finding northern spotted owls. The fact that it's an even-numbered year, when the birds typically have greater reproductive success than in odd-numbered years, has not helped. Four sites in the Nisqually River corridor were formerly reliable for spotted owls, but only two nests succeeded this year. One produced two young around Tahoma Creek and another two more in the park's Tatoosh Range in the Wahpenayo-Eagle Peak area. Observers reported a lone bird occasionally around Longmire. We hope to locate the Wahpenayo-Eagle Peak young to band, weigh, and take other measurements before releasing them.

Surveys at Mount Rainier began in 1983 and continue as part of a demographic study that stretches from Washington to California. Researchers conduct a minimum of three surveys per season on each of 35 spotted owl territories throughout the region. They want to know whether owls are living as individuals or in pairs, whether they are nesting, and if so, the stage of their nesting sequence. They also want to see if the owls are banded. Researchers place a colored band around the leg of every spotted owl they capture, which helps identify the bird if it's sighted again or recaptured. The nestlings, usually two, fledge in July. The adult birds get quieter and respond less often to recorded calls as the season progresses, making it more difficult to locate them and collect data. For a serious, long-time birder like me, the opportunity to observe a northern spotted owl at close range is like getting tickets to the Big Game. We take to the woods at the Nisqually River bridge near the Eagle Peak trailhead and follow an old prospector's footpath. We are soon walking over uneven ground without the benefit of a trail, clambering over logs and picking our way through the brush. The crew plays digital recordings of spotted owl calls; we listen intently for them to answer. No response. We walk toward a known nest site while playing more digital recordings. Still nothing. After more fruitless tries, a distant hoot returns, eerily reverberating through the forest. Without a word, the three crew members bolt off on a dead run in its direction. By the time I catch up to them, breathless, they've located an adult male northern spotted owl. It sits on a limb about 25 feet up a tree and about a hundred feet away. Known to be curious about humans and easily approachable, it stares at us. We gawk back. My binoculars help me study the bird, intermediate in size between a barn owl and a great horned owl. Chocolate overall with white spots, its dark brown eyes appear black in the dim forest. The poor light prevents me from determining the color of the bands on its legs.

Northern spotted owls do not build nests. They raise their young in treetops or in cavities of deformed or diseased trees with sufficient canopy cover for protection from the elements. They occasionally nest in abandoned raptor or squirrel nests or debris accumulations high above the forest floor. Spotted owls usually mate for life, are non-migratory, and defend their home range year-round. Its response to our digital calls is one form of territorial defense; it also attacks intruders with its claws. Nocturnal and most active in the two hours after sunset and before sunrise, it spends most of its waking hours foraging for food over an area that can exceed 25,000 acres.

Northern flying squirrels are the primary food source in this part of its range, and a pair of spotted owls can take up to 500 squirrels in a single year. Instead of actually flying, the squirrel glides between trees with the help of loose folds of skin stretching from foreleg to hindleg. It typically glides an average of 65 feet between trees, but some have been observed soaring 300 feet on a downslope. A flattened, wide tail serves as a rudder. Three to five young, each weighing no more than a U.S. quarter, are born blind and helpless in May or June. At around 40 days of age, they begin making brief trips from the nest into the forest.

At this junction of predator and prey, an intriguing set of interdependencies come together in a complex ecological web to benefit the owl, the squirrel, forest plants, and fungi.

Flying squirrels feed on truffles that they find underground from spring through fall. Truffles resemble potatoes and are the fruiting body of microscopic fungal filaments known as mycelia, which penetrate forest soils to form symbiotic relationships with plant roots. The fungus does not photosynthesize to produce its own food, so it finds nutrients through contact with the fine root tips of Douglas fir and over a hundred other forest plants. Plants produce carbohydrates via photosynthesis, and the fungus acquires them through a process called mycorrhizal symbiosis—from the Greek word meaning "fungus roots." In return, the fungus gives the plant access to additional nutrients and water.

Approximately 350 species of truffles occur in the Pacific Northwest, about 70 of which are sensitive or rare. The maturing truffle, which is how the fungus reproduces, emits an odor that attracts the northern flying squirrel and other woodland animals, including small mammals like voles, shrews, gophers, chipmunks, and larger mammals like deer, elk, bear, and mountain goat. All dig for the smelly but delicious truffle. When squirrels eat it, its reproductive spores remain undigested. If it defecates into a hole containing a Douglas fir root tip, the spores inoculate it. The spores germinate, mycorrhizae form and so begins the exchange of sugars and nutrients.

From the process's full cycle, the spotted owl and northern flying squirrel acquire essential foods. Moreover, the squirrel spreads fungal spores that germinate to form mycorrhizal relationships with plant rootlets that benefit both plant and fungus.

Once we catch our breath after locating the owl, the crew springs into action. I find a stick about three feet long, onto which one of the team members places a live mouse. I hold the stick and wait, trying not to worry about the bird's habit of clawing trespassers. Within moments, the owl drops from its perch and wings toward me. It grabs the hapless mouse in one graceful swoop, silently returns to its perch, and swallows its prey whole. The crew readies a second mouse for the end of the stick. This time, they think that the parent bird will carry it to its fledglings. Again, the bird flies straight for me, extends it talons, captures its prey, and makes its uphill getaway. Immediately, the crew hustles off, hoping to keep the bird within sight. After 15 minutes of strenuous off-trail travel, I catch up to them. They direct my attention to a nearby tree and there, just 12 feet off the ground, sit two juvenile northern spotted owls. Downy feathers cover their faces and breasts, but their wings and tails sport fresh flight feathers. Apparently unfazed by our commotion through the brush, they goggle down at us. These fledglings represent the species' prospects in this part of its range, but the odds are long for their chances of reaching adulthood. Juvenile birds often fall prey to great horned owls and northern goshawks. Others starve to death while struggling to find food after striking out on their own in the fall. If they make it to adulthood, survivorship is good, often up to 20 years.

Mount Rainier's spotted owl survey crews have changed their protocol since my field experience in 2012. In order to minimize the chance that the birds associate humans with food, researchers now place the "feeder mice" on a stump or downed log.

The crew begins gathering data while one of them assembles a pole-and-noose device to capture the young. He snares the first bird easily; we band and weigh it, take other measurements, then release it. The second bird is harder to corner but eventually we trap, band, and weigh it. Someone asks whether I'd like to set the bird free. I hold its feet together with one hand and its wings against its body with the other. Its gaze locks onto mine. Those dark eyes penetrate mine for a long, quiet moment as I ponder this icon of the Pacific Northwest old growth forests. Heeding the command, "Just let 'er go," I raise my arms, release my grip, and watch as the bird flies to a nearby perch. We finish our field notes and eat our mid-morning snacks while the two young birds sit nearby, seemingly oblivious of our presence. Our morning's work is over, but the task of ensuring the survival of the northern spotted owl has a long way to go.

Between the early 1970s when the northern spotted owl was first studied in the Pacific Northwest, until loggers had cut 5.6 billion board feet of mostly old growth in Oregon and Washington in 1987, the owl became the focal point of an intense battle between environmentalists and timber interests. The controversy arose because spotted owls prefer the late-successional old growth forests so highly prized by the logging industry, and habitat diminished drastically with their clearing. Forced to live on small remnants or other less-than-optimal patches, spotted owl numbers plummeted.

Biologists and environmentalists won a huge victory in 1990 when the U.S. Fish and Wildlife Service (USFWS) declared the northern spotted owl threatened under the U.S. Endangered Species Act because of habitat loss due primarily to logging. Prior to the decision, more than a million acres of old growth had been cut. Conversion of land for human use and natural disturbances like fire and windstorms added to the total. Biologists estimated that its habitat had shrunk by over 60 percent since European Americans arrived in the Pacific Northwest less than 200 years ago.

In 1994, the Northwest Forest Plan became the interagency blueprint to protect all plants and animals on federal lands within the range of the northern spotted owl. Its main thrust was to cut less old growth and instead harvest from those areas less preferred by the owls. The plan also contained strategies to mitigate the catastrophic economic setbacks felt by communities and families affected by the ruling. It sought to keep 20 percent of the remaining mature forests (those around 80 to 200 years old) and old growth forests (those older than 175 to 200 years) available for harvest.

The plan's authors hoped this would cushion the economic blow while protecting the remaining 80 percent of forested public lands. Sentiment outside of the timber industry and logging communities hotly opposed *any* old growth cutting, however, and harvest quotas fell far below the original predictions. The plan stemmed the tide of lost old growth and set aside 24 million acres of spotted owl habitat on public lands. Built on the notion that saving habitat in the form of vast land preserves would allow the population to rebound, it was a plan that looked good on paper but fell short of being a comprehensive recovery plan. The owl continued its decline.

Hoping to stop or reverse the bird's nose-diving numbers, the USFWS released the Spotted Owl Recovery Plan in 2008. The plan received immediate and overwhelming criticism, forcing a revision. The 2011 Revised Recovery Plan detailed the bird's continuing downturn, particularly in the northern portion of its range. Researchers and concerned groups realized that they needed to do more to secure its future. In the meantime, another threat had arrived on the scene, an interloper whose impact is still being felt in spotted owl country.

The threat is the barred owl. Twenty percent larger than the northern spotted, the barred owl requires smaller territories, adapts to more varied habitats, has a wide-ranging diet, nests earlier and more often, and can produce more young. These advantages have enabled it to replace spotted owls throughout its range. The loss of habitat posed the first major threat, but the barred owl brings a second significant peril.

Barred owls originally inhabited only the eastern United States, and some biologists believed that the treeless Great Plains restricted the bird's range. It is not clear whether its move west around the beginning of the twentieth century resulted from natural range expansion, human-caused conditions, or both. Forest fire suppression that allowed trees to grow larger and provide better nest sites, and the planting of shelterbelts around crops and settlements may have contributed to a range expansion north and west through Canada's boreal forest into the Pacific Northwest.

After listing the northern spotted owl as threatened, scientists became even more alarmed when data showed their continuing downswing. In a 28-year period in 11 study areas in Washington, Oregon, and California, they declined by 3.8 percent per year. During the same period, the overall population drop in the Washington study areas ranged between 55 and 77 percent. Biologists believe that competition with barred owls may be the primary cause of the shrinking spotted owl population. They also think that the barred poses a serious threat at Mount Rainier and that within this century the continued rate of decline could decimate northern spotted owl populations in this part of its range. Faring even worse in Canada, it is estimated that no more than six spotted owls remain in British Columbia and that the bird "faces imminent extirpation."

Both the 2008 and 2011 recovery plans recommended barred owl removal from northern spotted owl habitat, yet Fish and Wildlife Service staff agonized for more than seven years until ruling for the experimental removal of barred owls in four areas of the spotted owl's range. In 2013, a pilot program of eliminating barred owls—with 12-gauge shotguns—began in northern California's Redwood Creek area. Scientists soon noticed that spotted owls were returning to the test plots.

Killing one animal to save another usually generates controversy, and an animal rights advocacy group promptly sued the USFWS to stop the project. There is growing sentiment among researchers, however, that the northern spotted owl's dire situation makes the option viable. The service estimates that it needs to remove barred owls for at least four years in order to collect enough data to determine whether the program improves spotted owl population trends. Pilot areas include one in California, two in Oregon, and one in central Washington. They comprise about 2 percent of the total northern spotted owl habitat. Researchers plan to kill about 3,600 barred owls. If successful, the program will remove them from less than one-twentieth of one percent of their entire range. Whether the agency continues or extends the removals will depend on the return of spotteds to the target areas.

As the northern spotted owl continues its decline, the goal of stopping or reversing the trend rides on the habitat set aside by the Northwest Forest Plan and the success of the barred owl removal program. Adding to the challenge is climate change, which may play a decisive role in the owl's future. The cycle of fire, insect infestations, and disease outbreaks may shift because of climate change, and this in turn may influence prey or habitat. As biologists continue working to save the inquisitive owl, it's unclear just what its future holds. One thing remains certain, though. If we lose the northern spotted owl, I believe we will lose the soul of the forest.

Lava Flows and Age-Old Mountains

Longmire's natural history includes more than disturbance events, old growth forests, and iconic organisms. Smoldering lava flows that sideswiped gargantuan alpine glaciers and a mountain range millions of years older than Mount Rainier itself help complete the area's story. At 6:14 a.m. on the 24-hour clock (370,000 years ago), a slow motion collision between lava and an imposing glacier changed the area's geography. The colossal Nisqually Glacier towered hundreds of feet thick at present-day Longmire and extended many miles down valley. The Rampart Ridge lava flow began when gooey lava crawled from the mountain's interior to form the 4,000 to 5,000 foot high ridge that borders Longmire's west side. Standing in front of the National Park Inn and facing north toward the mountain, Rampart Ridge is on the left. The lava that became the ridge met the glacier's edge in what must have been a messy mix of hissing, boiling, and steaming as glacial ice guided and diverted the lava. You can still see the contact points where it flowed against snow or ice. On a reasonably clear day, a pair of binoculars helps locate rocky cliffs made of columnar andesite on the ridge. As the lava cooled slowly and uniformly from the bottom up, it contracted and cracked to form hexagonal patterns. Other flows in this part of the watershed that show contact with glacial ice include Ricksecker Point and Mazama Ridge.

Alert hikers traveling up Rampart Ridge can also find evidence of the area's recent fire history. Resulting from lightning strikes, fires helped create the conditions for a ridgetop plant community that differs from that in Longmire's lowlands. Visitors can hike or snowshoe the trail ascending Rampart Ridge year-round, but the summer months afford the best chance to observe the effects of fire that periodically course through the forest. Summer is also a choice time to appreciate the understory plants unique to drier and higher elevations. Ridge-bound hikers in spring and summer may enjoy the soundtrack of Pacific-slope flycatchers and Pacific wrens, two common birds more easily heard than seen. The flycatcher's repetitive *see-you-EET'* works as a steady counterpoint to the wren's complex and highly variable song. The tiny wren, scarcely larger than a thumb, seems to use its entire body to render a rolling tumble of notes that last up to 10 seconds.

Hikers move from the old growth in Longmire's western hemlock zone into a forest of pole-sized hemlock and Douglas fir aged at about 150 years old. Silver fir joins them, marking the transition into the silver fir zone that extends from around 3,000 to 4,500 feet in elevation. The silvery undersides of its needles give the tree its name, and the part of its scientific name that means "lovely" (*amabilis*) captures its beauty. These three continue upslope as the dominant trees in the mid-elevation forests. Yellow cedar, western white pine, and noble fir occur on moister soils; lodgepole pine and Engelmann spruce also occur but are rare in the park.

Joining the silver fir seedlings in the understory are bear grass and salal, mainstays where fire irregularly pulses through plant communities. Bear grass's immense root crown enables it to resume growing shortly after a fire. Salal also responds quickly, with its underground plant stems called rhizomes that store food and search out new areas. Indigenous people ate salal berries fresh or dried in cakes and used its leathery leaves to insulate pit ovens. Both plants are well adapted to long, dry spells, minimizing moisture loss during summer's drought-like conditions. Bear grass's long, narrow leaves radiating from its center help regulate temperature and reduce water loss. Salal's waxy evergreen leaves do the same. Its shiny, oval leaves and abundant blue-to-dark purple berries that ripen in midsummer make this common shrub easy to spot.

On top of Rampart Ridge, large trees are sparse or absent altogether, giving way to plants adapted to harsh conditions. Oregon grape, red huckleberry, and pipsissewa, all able to withstand drought and spring back quickly after fire, join salal and bear grass. Pipsissewa, or prince's pine, is a dwarf evergreen with many medicinal uses, including as a general tonic, or mixed with cascara bark in a tea to treat tuberculosis. These low-growing perennials and shrubs frame two spectacular views of the mountain.

For the best views of Mount Rainier from the Longmire area, hikers traverse the 3.5 miles and many switchbacks to the Eagle Peak saddle. The trailhead sits behind the Longmire residential area, nestled between the Nisqually River bridge and the Longmire Community Building. Just

Early photo of the Tatoosh Range by Lloyd Garrison Linkletter, noted Seattle-area photographer. Louise Lake is mid-ground left and Reflection Lakes above it and to the right. The Stevens Canyon Road, which bisects the area today, was not completed until 1957.
Courtesy of Mount Rainier National Park Archives

below the summit, the high meadow in full summer bloom boasts a wonderful array of subalpine wildflowers. Red heather, arctic lupine, hellebore, red columbine, common red paintbrush, tiger lily, and others all vie for attention along the trail. Woody plants include false box, Sitka mountain ash, subalpine fir, mountain hemlock, and yellow cedar. The stunning view from the top looks northeast down onto the Nisqually Glacier and the Paradise area. Looking south on a clear day, hikers can enjoy splendid views of Mount Adams, Mount St. Helens, and Oregon's Mount Hood.

Eagle Peak rises above Longmire to nearly 6,000 feet, anchoring the westernmost end of the Tatoosh Range. These rough, jagged mountains stretch skyward to almost 7,000 feet, forming a meandering east-west chain of about a dozen crags. Running along the park's south-central border, they are geologically distinct from and much older than 500,000-year-old Mount Rainier. Eagle Peak is part of the 22- to 26-million-year-old Stevens Ridge Formation, one of the mountain's geologic forebears. The remaining summits formed between 14 and 26 million years ago when three moderate-sized rock masses called plutons formed as magma and then cooled beneath Earth's surface. Compared to the geologically youthful Tahoma, the Tatoosh Range endures as the old-timer of Mount Rainier National Park, survivor of the millions of years of grinding erosion shaping these crumbling, foreboding peaks.

5

The Puyallup River:
Watershed under Pressure

Mount Rainier from the Thea Foss Waterway on Tacoma's Commencement Bay. *Drawing by Lucia Harrison*

THE RIVER'S END

Shades of gray dominate the color palette this fall morning on Commencement Bay. A dense curtain of ashen clouds obscures the upper reaches of Mount Rainier, 45 miles to the southeast. To the east and north, the cottony clouds appear soiled, so low they seem to snag the ground. Even the water has a leaden look, dull and lifeless. This is grayscale in nature: Pacific Northwest, November-style.

On Commencement Bay amidst industry and commerce, traffic and bustle, the Puyallup River completes its run to the Tacoma waterfront. It is born loud and wild, nearly vertical, high on the mountain's western flanks where the Puyallup and Tahoma Glaciers give up their waters. By the time the river reaches the lowlands, it flows are channeled and straightened, a subdued prisoner of urban development. Sitting on Puget Sound, Commencement Bay is a centerpiece of waterfront industry and development, hosting one of the West Coast's busiest ports. The city of Tacoma tumbles down the hillsides to meet the bay on all three sides; Browns Point on the east and Point Defiance on the west enclose it.

I spend the morning aboard a pollution patrol boat managed by the Citizens for a Healthy Bay, a local nonprofit. The only one of its kind in the south Sound, the boat motors nearly 2,000 miles making 120 trips a year on the bay. Most of its work is to maintain visibility, much like how a state trooper's presence on the side of the road slows traffic. When a spill or other situation arises, it is part of an early response warning system.

We follow a tug escorting a giant barge loaded with scrap iron. Great blue herons sit atop pilings like sentries. A flotilla of Barrow's goldeneyes

The Port of Tacoma, one of the busiest on the West Coast, lies on Commencement Bay at the mouth of the Puyallup River. *Port of Tacoma and Northwest Seaport Alliance*

reluctantly breaks rank to let us pass. Longer than a football field, a container ship from Singapore is simultaneously loaded and unloaded. Our boat's radio crackles as pilots and dispatchers negotiate traffic and adjust shipping schedules. Up and down the waterways, hard-hatted dockworkers in Day-Glo vests wrangle pipe and wrestle rigging. The waterfront buzzes with activity. This is the Port of Tacoma on Commencement Bay, catchment basin for the Puyallup River, in the twenty-first century.

But the footprints of human presence in the lower Puyallup watershed have come at a great cost. A series of ecological impacts have cost millions of dollars and taken over 30 years to remediate. Crews have removed more than two million cubic yards of contaminated sediment from the bay, enough muck to fill a line of dump trucks from Tacoma to Fresno, California.

Not even a sailor drunk on grog could have conjured present-day Commencement Bay. During the spring of 1841, the task of the U.S. Naval Exploring Expedition was similar to British Captain George Vancouver's and his crew 49 years earlier: explore, survey, and map the inland waters of the Salish Sea. Vancouver named the southern section Puget's Sound for Second Lieutenant Peter Puget, who with Ship's Master Joseph Whidbey led the weeklong surveying party in the south Sound. They sailed only the west side of Vashon Island, however, along today's Colvos Passage, and did not enter the large bay into which the Puyallup and many smaller watercourses flowed.

When Lieutenant Charles Wilkes and his crew aboard the *Vincennes* and the *Porpoise* entered Puget's Sound in 1841, they were already three years into the U.S. Navy expedition that had taken them to the Antarctic and the South Pacific. When it ended the next year, their written records and thousands of scientific specimens made such a significant contribution to the development of science in the United States that the Smithsonian Institution became the repository for the expedition's collections. Wilkes's task was as straightforward as Vancouver's, but with an additional twist. He was to chart the territory—in this case, for potential settlement by U.S. citizens—but also to give the British, with whom the Americans tussled for possession of the Oregon Territory, a tacit yet clear message of the intent of the United States to settle the Pacific Northwest. A time-tested method for conveying that desire was to name the waterways, islands, and prominent points of land. Sometimes using the physical description of a feature, at other times honoring men and ships from the U.S. naval tradition, and at still others using the surnames of his crew members, Wilkes bestowed a record of almost 300 place names on Pacific Northwest locations. He began by charting the protected waters that he called Commencement Bay.

Looking at the bay today, it's hard to imagine the scene that Wilkes and crew encountered. Though spare, their journal entries paint a rough picture. On entering the bay, Wilkes found,

> An indentation on the southeast, three miles deep by two miles wide. The water is too deep for anchorage, except around the borders of the bay, on its south shore. The head of the beach has an extensive mud-flat where several small streams enter, among them the Puyallup....The Puyallup forms a delta, and none of the branches into which it is divided are large enough for the entrance of a boat.

Surveyors later estimated that the network of tidal mudflats and wetlands covered between seven and nine square miles, comprising a biologically rich estuarine system similar to the Nisqually estuary about 20 miles to the southwest. One crew member recorded the bay as "a fine sheet of water...surrounded by a range of low hills covered with splendid trees."

First People, First Stewards

Fair weather prevailed during Wilkes's visit in May 1841, allowing views of the mountain that the bay's first people knew as *Takkobad (Tuh-ko'-bud)*. Their Puyallup descendants sometimes refer to it as "Our mother our beloved Mount Takoma." Of the people that Wilkes saw in the newly named bay, he wrote that, "The Indians were at this season of the year to be found on almost all the points." Here at the water's edge, where the Puyallup River enters the bay and the south Sound, native people had lived for millennia.

Linguists believe that "Puyallup" originates from the Lushootseed word *s'puyalupubsh*, meaning "people of the bend." One look at a nineteenth century map of the river as it wound into Commencement Bay explains why. Prior to channelization and realignment intended to reduce flooding, the lower Puyallup was a sinuous waterway. Noted tribal member Henry Sicade offered another definition. In 1916, he told the *Tacoma News* that it meant "generous people," from *Pough*, which means people, and *allup*, meaning generous. The tribe's contemporary interpretation is "the people of the generous and welcoming behavior to all people (friends and strangers) who enter our lands."

For her 1935 ethnography *The Puyallup-Nisqually*, anthropologist Marian W. Smith interviewed seven Puyallup tribal members ages 68 to over 80. Their stories of the old days helped her pinpoint the location of ancient villages that stretched southeastward from the Puyallup River mouth. Drawing on their memories and the stories of their elders, Smith assembled an extensive account of Puyallup-Nisqually lifeways and Puyallup village locations.

Since time immemorial, four villages thrived at the mouth of the river in what is today Tacoma's downtown waterfront. The "large and important" village known as *Puyallup* near the present intersection of 15th Street and Pacific Avenue was home to people Smith's informants called "real" Puyallups. Three villages ran contiguous with it, the second at present-day 24th Street and Pacific Avenue. Its Lushootseed name translates to "the place where the tide has gone out." *Catcqad*, dubbed a main village by other anthropologists, sat on the river's north bank near the old Cushman School, which later became the Cushman Hospital. Interstate 5 passes over the river at the age-old site.

North of these villages lining the west side of Commencement Bay, "was a continuous line of settlements from the present Tacoma Hotel [today's 10th and A Streets] to the Stadium [today's Stadium Bowl]." An ancient Puyallup burial ground lay between the settlements and Point Defiance. When a settler asked permission to clear it for cattle grazing in 1882, the canoes of the dead still hung suspended from the trees. He received permission and the help of Indian students from Cushman School to clear the site. Over 50 years later, one of the students told his story to Marian Smith.

Villages on the east side of the bay included those at the mouth of Wapato Creek near the end of today's Hylebos Waterway, on the Hylebos itself, and at Browns Point, the bay's easternmost promontory. The Browns Point area was called "hidden water" in Lushootseed, presumably because it hides the view of Puget Sound farther north.

Ancestral Puyallup villages extended even beyond the Commencement Bay area. Oral histories have long told of settlements across the water on the Kitsap Peninsula and on Vashon Island. A large village at the mouth of a creek in Gig Harbor gave the Puyallups control of the only water entrance to the southern waters of the Salish Sea.

In 1996, an archaeological excavation at Quartermaster Harbor on Vashon Island found evidence that ancestors of the Puyallup Tribe had lived there in large winter villages over the last thousand years. They called the inner part of the harbor *sdₐgʷalaɫ,* which means "enclosed water." Elders knew the site as a Lushootseed word for madrones, for the trees lining the shoreline before erosion swept them away. Known to archaeologists as the Burton Acres Midden Site, the people living there left behind thousands of shells, fish remains, and other detritus. They were the *S'Homamish,* Puyallup predecessors. People of the saltwater, their name translates to "swift water dwellers." They were expert canoe people who masterfully navigated the often rough and choppy waters of the south Puget Sound. Findings at the site showed that they harvested and dried seasonally abundant fish, especially migrating herring, and collected shellfish, especially littleneck and butter clams.

Back on the mainland, villages extended about 15 miles upriver. The name of the village where Clark's Creek empties into the Puyallup at today's 66th Avenue East translates to "raven." Another village, *T'kwawkwamish,* sat at the confluence of Voight Creek and the Carbon River, a Puyallup River tributary. The village's name refers to the upper reaches of the Carbon River, signifying the distinction of the upper drainage Puyallup people from the saltwater people living around Commencement Bay. Along with the two settlements flanking *T'kwawkwamish,* these were the furthest upriver Puyallup villages. Like the Nisqually upriver villages, these villages had a Sahaptin language influence, suggesting a gradual but steady movement of people from central and eastern Washington into the Salish Sea basin.

There is no evidence of permanent, year-round villages beyond *T'kwawkwamish,* but several sites within the boundaries of Mount Rainier National Park indicate the seasonal presence of Native Americans. Two projectile points recovered not far from the Wonderland Trail between Klapatche and St. Andrews Parks, the highest elevation where artifacts have been found in the park, suggest hunting activities. Another projectile point and tool fragments were found in the Golden Lakes area. It is possible that these people were of Puyallup ancestry who lived in downriver villages.

The Bay: Developing, Degrading, and Detoxifying

Within the decade after Lieutenant Wilkes and his crew had completed their voyage, the first European American settlers began staking claims in the Commencement Bay area. A small water-powered sawmill, a portent of things to come, began operating in 1852. After a larger mill opened in 1869, the first shiploads of lumber departed the bay. The push for development intensified when the Northern Pacific Railroad selected Tacoma as the terminus for its transcontinental railroad in 1873. During the boom years of the 1880s, developers filled in portions of the bay to create industrial sites, making way for the first building on the tideflats, the world's largest sawmill. These were prosperous times in Tacoma, and boosters began calling it "The City of Destiny." The outlook for Commencement Bay's health, however, looked grim.

The beginning of the twentieth century brought the development of dockage areas on the bay. Eager entrepreneurs created the City Waterway (later renamed the Thea Foss Waterway for the woman who started a rowboat rental business in 1889 that became Foss Maritime). Dredging began on the Hylebos Waterway. Sawmills now lined the shore, giving the city the title of "Lumber Capital of the World." Nearly two dozen shipping lines carried lumber and grain to ports worldwide. The Port of Tacoma, established in 1918, purchased 340 acres of tideflats for development. Scarcely more than 10 years later, the port filled an expanse of mudflats and salt marsh twice that size to create other shipping waterways. Wood waste from the mills, mixed with smelter slag containing arsenic and other toxic metals, became the fill material for roadways and industrial sites. Chemical plants and foundries released toxic organic chemicals and dangerous metals into the bay's waters, the groundwater, and the air. Refineries and tank farms leaked petroleum products from underground storage tanks and pipelines.

Commencement Bay's degradation continued unabated and unregulated for one hundred years. Two factors contributed to its decline. First, there were no regulations in effect requiring industries to comply with protective environmental standards. Without established safeguards for the handling and disposal of chemical wastes or guidelines to follow in the event of spills, businesses dealt with hazardous materials in any way they saw fit, which resulted in the haphazard and widespread discharge of toxic substances. Compounding the problem was a long-held misconception that tidal waters carried away pollutants and replaced them with cleaner, pollution-free water. Businesses dumped waste and the city discharged sewage into the bay, transforming the inner waters of Commencement Bay into a poisonous soup that included cancer-causing polychlorinated biphenyls (PCBs), industrial solvents, petroleum products, and a variety of toxic metals. It made a lethal chemical cocktail, pushing the bay to its breaking point.

In the late 1970s, studies by the National Oceanic and Atmospheric Administration (NOAA) showed a correlation between chemical contaminants in sediments and the organisms living there. Rather than being carried away by the tides, the pollutants accumulated in the sandy, muddy waterway sediments and tideflat soils. The toxins found their way into the resident organisms and became concentrated as they worked their way up the food chain. They manifested as liver disease and cancerous tumors in English sole and rock sole, common bottomfish. Additional research by the Washington State Department of Ecology (DOE) confirmed that the bay's sediments were highly contaminated.

Commencement Bay was just one of many sites across the the country in need of cleanup. Federal legislation in 1980 had authorized the Environmental Protection Agency (EPA) to respond to the nation's pollution epidemic by creating a list of polluted sites and requiring a cleanup program for each location. The Commencement Bay Nearshore/Tideflats project joined the Superfund National Priorities List in 1983. Under the cleanup plan, the EPA took the lead in dealing with polluted sediments while Washington's DOE oversaw the identification and control of pollution originating from specific sources. DOE staff interviewed facility owners and learned about the historical uses of the properties. They sampled air, sediments, soil, and water. They helped business owners improve how they handled and disposed of dangerous chemicals. Once the EPA reduced or stopped the flow of toxic releases, they treated the polluted sediments, covering some with a layer of clean sediment and capping them in place. Other sediments were removed and placed in onshore landfills. Completed in 2005, the removal totaled two million cubic yards, the largest sediment cleanup in Superfund history.

Progress was steady and groundbreaking. The Simpson Tacoma Kraft pulp and paper mill was the first intertidal cleanup completed under the Superfund. Remediation of the St. Paul Waterway resulted in its removal from the list in 1996, making Commencement Bay the first site to de-list a waterway.

Other groups came forward to join the cleanup. The Port of Tacoma created Gog-le-hi-te Wetlands Park by converting a former city landfill to a functional wetland. The City of Tacoma pioneered a major cleanup on the Thea Foss and Wheeler-Osgood Waterways. It footed over half of the project's $105 million tab, collaborating with state agencies and private industries to restore habitat at multiple locations along the waterways and tideflats. The Commencement Bay Natural Resource Trustees, comprised of federal and state agencies, the Puyallup Tribe of Indians, and the Muckleshoot Indian Tribe, funded habitat restoration at 10 sites on the bay. The Port of Tacoma, the Tacoma-Pierce County Health Department and the cities of Fife, Fircrest, Ruston, and Tacoma assisted in pollution prevention, enforcement, and cleanup.

Nearly 40 years later, restoration continues. According to Sheri Tonn, Pacific Lutheran University chemistry professor and inveterate activist in the Commencement Bay saga, recontamination looms as a critical issue. Twenty years after the completion of a cleanup project on Hylebos Creek, arsenic and other heavy metals leaked into the water. Industrial polluters aren't the only source of contamination, either. Nonpoint source pollution continues as a substantial, largely untraceable threat. Soapy water from curbside car washing, fertilizer and pesticide runoff from lawn care, or vehicles dripping fluids or brake dust onto city streets all find their way into the bay through the city's 18,000 storm drains and 500 miles of storm sewers. Nearly 20 percent of Tacoma's stormwater enters Commencement Bay, guaranteeing that every time it rains, toxic substances wash into storm sewers to recontaminate it.

Finally, work at some sites proceeds slowly, and at others, not at all. The Occidental Chemical Corporation, first mentioned in a *New York Times* article in 1981, spilled or dumped between four and eight million pounds of hazardous materials into the ground or into the water between 1929 and 2005. Toxic dry cleaning fluids, PCBs, and other dangerous chemicals have created a deadly mix in the groundwater extending 160 feet below sea level and beyond the plant's footprint. Up to two million pounds of poisons may still be on site or in the groundwater plume. Cleanup and restoration is still years away. "We have 12,000 data points on the plume," said Tonn. "We know where it is, we just don't know what to do about it yet."

On a scale of one to ten where one symbolizes the worst of the Commencement Bay pollution problem and ten represents water so clean that you could safely eat the bottomfish, the bay's health currently registers at a five or a six on the scale. It remains a Superfund site. While still not a healthy place for all marine species, the bay is becoming less contaminated. The legacy of development without caution, says Tonn, is the message for the future. As a founder of the Citizens for a Healthy Bay, she believes that development and conservation must coexist. With environmental standards in place, stiff penalties for polluters, a dogged focus on stormwater, and heightened public awareness, the scene is set for continued improvement.

MORE TROUBLED WATERS

At the watershed's upper end, the Puyallup bears no resemblance to its downriver self. Multiple meltwater fingers drain the glacier to birth the North Puyallup River in an area so steep and rugged that I can only survey it through binoculars from a mile away. Andesite monoliths form a towering backdrop, shrouded in gauzy fog one moment and bare the next. The rocks stand perpendicular and impenetrable, vanguards of the Puyallup Glacier. Intrepid bands of conifers edge up the lower slopes to cling precariously onto narrow rock ledges. On one of the ledges, a nanny goat threads a sinuous path for her yearling. She takes a few cautious steps, pauses, and looks back. The hesitant kid paws the ground and surveys the rockbound route. Mother waits. Finally, in a graceful burst of youthful energy, it bounces across the craggy passage.

A lobe of glacier sits just around the shoulder of the most massive rock formation, the river of ice plunging to its terminus. Through binoculars, I make out a yawning ice cave at the glacier's snout. From it, the newborn river hurtles wildly down a precipitous rock face and rumbles through this narrow, steep-sided canyon. The Wonderland Trail intersects the North Puyallup here, where wildness rules all. The color of dirty mop water, the river rushes madly, hops boulders, plummets 15 feet, and then another 30 into a swirling, turbulent pool below the bridge deck. Standing weak-kneed on the sturdiest of bridges, I look down on mammoth boulders hauled out like giant sea lions waiting for the next high water to sweep them downriver. Huge logs with rusty spikes driven into them, remnants of a previous bridge, lie strewn about like matchsticks. Gnarled rebar protrudes from concrete fragments. The surge of air down the channel creates an eerie, menacing breeze. And it is LOUD. The river thunders past, a steady drum roll punctuated by occasional boulders that hammer syncopated bass notes against its bed, charging on with great purpose, angling northwest to just outside the park boundary. There it joins with the South Puyallup River, which flows from the adjacent Tahoma Glacier. The Tahoma Glacier's moraines range in color from charcoal to russet to brown to a lovely golden, all the artwork of minerals. The water often runs a yellowish-brown café au lait. With the glacial waters of the mountain's western face now conjoined, the Puyallup rolls on toward Commencement Bay. Soon it will take on its placid, down valley appearance.

Three major rivers comprise the Puyallup watershed. The Carbon and White Rivers join the Puyallup as tributaries, creating a watershed with

728 miles of rivers and streams that drain an area of over 1,000 square miles. Seven times the area of Seattle, the watershed transports 60 percent of Mount Rainier's waters to Puget Sound.

Were it not for the discovery of coal deposits in the area, the Carbon River might have retained the name given it by Lieutenant Wilkes, the Upthascap River. It originates at the Carbon Glacier's terminus and with its 19 tributaries makes up about 30 percent of the Puyallup's flow, ending its 33-mile run by emptying into the Puyallup just outside the Orting city limits.

The White River flows from the Emmons Glacier on Mount Rainier's northeast side, running nearly 70 miles to its confluence with the Puyallup River near Sumner. The West Fork White River drains the Winthrop Glacier basin, just west of the Emmons, and empties into the White. High above the valley floors, these adjacent glaciers grind relentlessly at the bedrock beneath them, producing a silt so fine that the river transports it for its entire run. The river draws its name from these suspended sediments—known as glacial flour—that give the river its milky color.

The White River drains almost half of the Puyallup watershed, but has emptied into the Puyallup only since the early 1900s. The Osceola Mudflow had cut a channel that emptied the White into the Green River and into Puget Sound, until flooding forced the Army Corps of Engineers to consider its options. To control flooding, the corps diverted the White River into the Stuck River, which flows into the Puyallup.

The bounty of the Puyallup watershed that sustained Native Americans also drew the first European Americans to the area. The dense and mighty forests, the abundant salmon and other game, and the promise of fertile agricultural lands provided great incentives to the newcomers. With one crucial exception, the same conditions existed at the nearby Nisqually River estuary. Plentiful timber and game abounded there, but the shallow waters of the Nisqually Reach prevented easy access for large ships. Commencement Bay's deep waters made for convenient navigation for the largest ships of the time. The bay's potential as a seaport sealed the Puyallup's fate as the river drainage with the greatest influx of settlers and ultimately, the greatest environmental impacts.

Drawn by one of the largest seaports on the West Coast and thriving industry, people continue streaming into the lower Puyallup basin. A

10 percent population increase between 2000 and 2010 made it the third most populous watershed in the state. Developing communities naturally strain local water resources. Besides greater volumes of wastewater in municipal systems, new septic systems in outlying townships raise the risk of leaks and groundwater contamination. The biggest threat to Puyallup basin water quality, though, is the dramatic increase in impervious surfaces. Between 1986 and 2006, roofs, roadways, parking lots, and other nonporous surfaces increased by 47 percent, equivalent to 35 square miles of asphalt—enough to pave more than half of Tacoma. These hard surfaces prevent water from entering the ground directly, where the soil absorbs contaminants and other chemical impurities, and replenishes groundwater. Without the cleansing properties of soils, surface water flows freely, carrying pollutants into creeks, streams, or the stormwater system.

Studies show a strong correlation between impervious surfaces and degraded stream health. Where impervious surfaces cover between 10 and 20 percent of the ground, reduced water quality, higher flooding risks, and other undesirable water-related conditions occur. The lower Puyallup watershed has already eclipsed that threshold, with impervious surfaces covering between 11 and 46 percent of the ground.

Fecal coliform, dissolved oxygen, and elevated water temperatures represent the greatest threats to Puyallup basin water quality. Fecal coliform is a common, naturally occurring bacteria in the digestive tracts of humans and other warm-blooded animals. Most of these bacteria are benign, but they are an indicator of more dangerous water-borne pathogens, such as those causing dysentery, cholera, and other diseases. Problems arise when fecal coliform accumulates in water. Faulty wastewater treatment facilities or leaky septic systems can release fecal coliform into local waters. Farm animal waste washes into and contaminates waterways in agricultural areas. Careless pet owners who fail to dispose of wastes properly contribute to fecal coliform levels in stormwater and surface water systems. When the levels rise above acceptable limits set by the EPA, the agency considers the water polluted and requires remediation.

Dissolved oxygen (DO) and water temperature are inversely related indicators of water quality. Dissolved oxygen indicates the amount of oxygen present in water that is available for respiration for fish and other aquatic life. Higher levels provide better conditions for organisms than low levels. Cold water holds more oxygen than warm water, so DO tends to be higher in winter and early spring. Warmer water temperatures equate to a lower DO concentration. Water temperature is especially important to salmon and trout, which need a range between 48° and 64° Fahrenheit to live. Temperatures increase naturally during the summer months, but the removal of streamside vegetation, eroded stream banks, and polluted runoff drive them even higher.

The Washington State DOE and the Pierce County Public Works and Utilities Department each administer programs that monitor stream health. With problems that include fecal coliform, low DO, water temperature, and other issues, the DOE currently lists more than 60 miles of streams in the Puyallup watershed as "impaired waterways." None of the 41 sampling sites carries a water quality ranking of "excellent" or "good." Over half carry "poor" ratings, with the remainder listed as "fair."

Pierce County staff combine two indices to create a comprehensive stream health indicator, its annual Surface Water Report Card. They use the

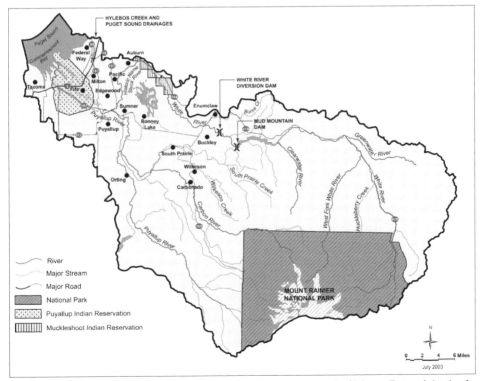

The Puyallup River watershed, the most populous and most developed of Mount Rainier's lowland basins. *Map data courtesy of King County, modified by Kirsten Wahlquist*

Water Quality Index to measure a broad range of the stream's physical and chemical properties that include fecal coliform, DO, water temperature, nitrogen and phosphorus levels, pH, turbidity, and total suspended solids.

The second measure, the Benthic Index of Biotic Integrity, uses organisms that live in the streams as indicators of stream health. Benthic macroinvertebrates that include worms, snails, freshwater clams, caddis-flies, stoneflies, and mayflies indicate the water's relative health. Their short life spans, small range, key role in aquatic food webs, and sensitivity to water clarity and pollution make them the "canary in the coal mine" of freshwater systems. Wherever benthic macroinvertebrates abound they indicate good water quality. Their absence or the presence of pollution tolerant species indicates lower water quality.

County personnel combine results from the Water Quality Index and the Benthic Index to yield a grade for Pierce County's lakes and streams. In the Puyallup River watershed, the overall stream grade from year to year fluctuates between C and C-.

FIGHT LIKE A SALMON

On a wet and unforgiving autumn day, I am on a scavenger hunt of sorts in the upper White River watershed. Mount Rainier National Park fisheries biologist Ben Wright tracks bull trout while I stumble through the brush after him, trying to keep up. The project's primary objective is to study the fish's passage through the Mud Mountain Dam. Today's work uses radio telemetry to locate fish. We also plan to find and record their breeding sites, known as redds. Ben wears a receiver and carries a handheld antenna, trying to pick up radio signals from one of 50 fish outfitted with radio collars. So far, the airwaves remain silent. While he continues trolling for signals, Ben leads us through the brush alongside the creek so that we don't disturb the fish.

The bull trout hasn't always enjoyed such positive attention. Anglers regarded them as salmon killers and egg stealers and threw them onto the riverbanks to die. Some state fisheries programs placed bounties on them. That has changed with a new understanding of the bull trout's importance as an apex predator with an integral role at the top of the food chain in aquatic ecosystems.

We arrive at the creek's confluence with the White River and begin hiking back up the creek toward our starting point. A rivulet just a few feet wide and several inches deep gives the bull trout everything it needs as it scuffles for survival and the opportunity to reproduce. It has the narrowest spectrum of habitat suitability of the region's trout and salmon, requiring cold, clean water and complex, interconnected habitats. This makes it an excellent indicator of healthy forest and aquatic ecosystems. Its limited tolerance to water temperatures may place it at high risk in the coming decades.

Ben's well-trained eye quickly spots a bull trout redd. These shallow impressions lined with gravels ranging in size from peas to ping pong balls are the next generation's nurseries. A mating pair guards one. Light spots freckle their dark bodies; a beautiful cream-colored contrast edges their pectoral fins. The female's tail appears worn from digging the redd; the male feistily wards off a competitor. He remains nearby to defend the area for up to two weeks after spawning. The fish become sexually mature at five to seven years of age and may breed multiple times, either annually or nonconsecutively.

Bull trout have complex life histories in which they live as either residents or migrants. Resident fish live in headwaters or high elevation streams like the one we're on today. They often spawn and overwinter in the same mile or two stretch of water. Biologists classify migratory bull trout as either fluvial, adfluvial, or anadromous. Fluvial fish travel more widely than residents, but remain in rivers and streams their entire lives. Adfluvial fish move between rivers, streams, and lakes. Anadromous fish are born in freshwater, migrate to saltwater, and later return to their birthplace to spawn. Anadromous bull trout at Mount Rainier migrate in spring down the White, Carbon, and Puyallup Rivers to Commencement Bay. Most do not travel extensive migration pathways, but some appear to enter the open ocean. They return in summer or fall to breed again in their natal waters. Remarkably, the recovery of some tagged fish shows that they're capable of switching strategies over the course of a lifetime, probably in response to the availability of food.

Bull trout join steelhead, coastal cutthroat, and the five species of Pacific salmon as the eight members of the Salmonidae family commonly found in the Puyallup River system. Except for bull trout, all reside in the genus *Oncorhynchus*, from the Greek words *onkos* for hook and *rhynchos* for nose. Even casual observers of spawning salmon are quick to understand the word's roots. With the exception of a few resident species, most salmonids are anadromous and die after spawning once. Only the bull trout, steelhead, and coastal cutthroat spawn multiple times. All five species of Pacific salmon inhabit both the Nisqually and Puyallup watersheds: Chinook, chum, coho, pink, and sockeye. No records remain of salmon populations from the 1800s, but biologists agree that today's fish battle for survival at a mere fraction of their original extent, having become locally extinct in over 40 percent of their range in less than 200 years. The leading causes of the declines include overharvest and habitat fragmentation, degradation, and loss.

In response to crashing salmonid populations, Washington's first fish hatchery began operating over 120 years ago. Today 12 federal, 83 state, and 51 tribal hatcheries produce over 75 percent of the salmon caught in Puget Sound that annually contribute $1 billion to the state's economy. Without hatchery production, it is likely that many populations would face local extinction or fall to levels low enough to force curtailment of harvests. Two state hatcheries in the Puyallup watershed, along with the Muckleshoot Tribe's White River hatchery, and the Puyallup Tribe's Diru Creek and Clark's Creek hatcheries, collaborate in fisheries enhancement. Together they annually rear and release millions of Chinook, chum, coho, and steelhead.

The most telling indicator of current salmonid health is the listing of bull trout and Puget Sound Chinook in 1999 and Puget Sound steelhead in 2007 as threatened under the U.S. Endangered Species Act (ESA). Protection under the act for the three species includes captive breeding in hatcheries, improving water quality and instream flow, harvest regulations, and habitat acquisition and restoration.

The Chinook is at the greatest risk of the three species, experiencing a 60 percent decline in the Salish Sea between 1984 and 2010. Designation of critical habitat helps some, as does the Puget Sound Chinook Salmon Recovery Plan. It advocates for action "across all H's": mitigating the effects of *hydropower* dams that impede migratory runs, *habitat* protection and restoration, *hatcheries* that supplement the numbers of wild fish, and closely monitoring their *harvest*.

Two runs of Chinook enter the Puyallup watershed—the White River spring and the White River summer/fall. The Muckleshoot Tribe's hatchery at Buckley is a focal point for Chinook recovery, annually producing about 350,000 fish. Returns have been encouraging in recent years, ranging from hundreds to over 16,000, the most ever recorded. Wild fish number less than one-fourth of the total run, suggesting that enhancement and supplementation remain vital to their survival. Fisheries biologists pin their hopes for the species on continued enhancement, conservation, habitat restoration, and on the fighting spirit of the mighty, resilient fish.

The lone exception to dwindling salmon populations is the highly adaptable and prolific pink salmon. The fastest growing salmon in the world enjoys the broadest spectrum of habitat suitability. Pinks are the Puyallup watershed's most abundant salmonid, with between 950,000 and 1.5 million returning in recent years. This wild stock thrives without hatchery support to maintain its two-year life cycle. Even-year pinks are uncommon in Washington but do occur. The pink's life cycle is less complex than other Pacific salmon as they move quickly through the Puget Sound to feed mostly on primary producers like plankton. Biologists are working to understand the sharp rise in the pink runs in the last decade. It is not clear whether the upsurge signals overall improvement of salmonid conditions, or—if pink numbers correlate with algal blooms, for example—it is a sign of the continuing decline of Puget Sound water quality.

CALL THEM KINGS

Conditions have favored this pocket of spring-run Chinook eggs on the Greenwater River, a White River tributary. Nestled in gravel that can be either an incubator or a graveyard, several hundred eggs have survived moderate winter floods and the sediment that often suffocates them. Pinkish-red, spherical, and about the size of a pencil eraser, they have endured and developed over these last few months. Eyes goggle eerily through soft, translucent shells. Developing organs are visible, too. At just the right time, the species' newest stock will emerge from their shells. Now called alevins and about an inch long, they continue feeding from the nutrient-rich yolk sac attached to their abdomens. They are not yet good swimmers, and so wriggle down into the gravel for protection where they remain a month or more until they completely absorb the yolk sac nutrients. Sometime between January and March, the sac is spent and the young salmon disperse to begin the dangerous business of fending for themselves.

Just an inch or two long, the fry have two imperatives: find food, and avoid becoming food for something else. Waterborne insects or terrestrial insect larvae fulfill the first requirement, stealth and luck the second. They seek cover under rocks or logs, or amid dense vegetation along the river's margin to avoid predation by birds or larger fish. Even in this vulnerable stage, they begin the downriver migration to Commencement Bay and beyond. From March through August more than three-fourths of the White River spring Chinook move downstream as fry.

The fry develop vertical markings on their flanks during their trip to saltwater that help them hide from predators. Up to six inches long now and known as parr, they're still highly vulnerable to predation. They'll soon undertake their next major transformation: from freshwater to saltwater organisms.

Internal rhythms and increasing day length trigger physiological changes that prepare the now-schooling smolt for saltwater lives. Chemical compounds in the fish's scales cause its vertical bars and dark spots to give way to the dark back and silvery belly that help it evade predators in the ocean's open waters. Gills and kidneys undergo changes that will enable the processing of seawater. The fish produces hemoglobin that will aid respiration in the saltwater's lower oxygen concentrations.

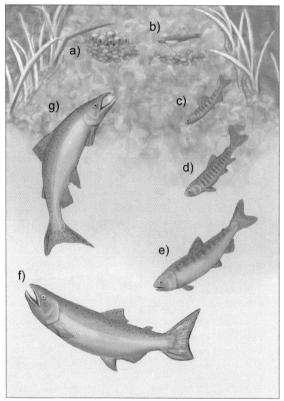

The Chinook salmon life cycle: a) The eggs develop on a gravel bed in the river over the winter; b) alevin with yolk sac still attached; c) fry; d) parr, with vertical marks for camouflage; e) smolt; f) oceangoing adult; g) spawning adult. *Kirsten Wahlquist*

While completing these physiological changes, the White River smolts continue their movement toward Commencement Bay. Swimming actively downstream at night to evade predators, most arrive by August. Some smaller fry or developing smolts may allow the current to carry them tail-first downriver.

At the river's mouth, the smolts linger in the food-rich estuaries. They feed heavily, bulking up to help ensure ocean survival. Estuaries also offer cover from predators and allow the fish to complete their transition to saltwater organisms. The next major challenge lies ahead: life at sea.

Mortality runs high among salmonids in the ocean; only about 5 percent survive marine life. The odds increase as the fish grows. Chinook fare somewhat better because of their larger size, but the first few months are the most dangerous for all salmon. Resident pods of orcas, lamprey, spiny dogfish, and seabirds all prey on Chinook in saltwater. Despite the risk, most remain at sea for two to four years, moving from the bay into Puget Sound and then beyond into the Pacific Ocean. One study found that most Chinook migrated less than 500 miles.

Abundant food supplies in the ocean make it easier for salmon to grow larger there than in rivers, so it benefits them to stay as long as possible. Their feeding rate and high metabolism contribute to rapid growth that may increase their weight a thousandfold or more during their time in saltwater, where most salmon gain 99 percent of their body weight. Seagoing Chinook eat nutritious prey high in calories, fat, protein, and unsaturated fatty acids that include small fishes, squid, larval crabs, krill, and other crustaceans. Mature, they range from 15 to 50 pounds and, as the largest of the Pacific salmon, bear the nickname "King." Some call a large one "Tyee" from the Chinook Jargon word for chief or king.

Usually between two and four years of age, the White River spring Chinook begin undergoing hormonal changes to prepare for the journey home. Feeding rates decline and then stop altogether. Reproductive organs begin developing. The osmoregulatory system, which once prepared the fish to transition from fresh to saltwater, now prepares it to return to its freshwater origins.

Living on the fat stored in their bodies, the salmon abandon their map and compass-like orientation system in favor of an olfaction-based homing system. They now navigate by odors learned as juveniles to locate their natal river to begin the journey home. Between 95 and 99 percent return to their birth rivers and streams. Those straying to rivers that did not birth them colonize new habitat and help increase genetic diversity. The fish are sexually mature and ready to spawn, but more challenges lie ahead before eggs and milt combine to create the next generation.

The fish return to the Salish Sea and Commencement Bay in early spring and begin working their way up the Puyallup River in April and May. The channeled and straightened lower section is the shortest distance between two points, but the absence of large woody debris makes it difficult to find cover for resting periods. They continue up the Puyallup, enter the White River, and in June begin arriving at the Buckley fish barrier. Without a fish ladder at the Mud Mountain Dam five miles upriver, the barrier funnels the fish into a holding pen. There they await transfer into a tanker truck that takes them to a re-entry point on the river above the dam. For these salmon born in the tributaries of the upper White, they are almost home. One final challenge remains—the act of spawning.

The female arrives first on the spawning grounds to select and prepare a redd in which to deposit her eggs. She must consider water depth, velocity, and gravel type. Competition for the best sites is fierce and each must guard and vigorously defend her choice. Facing upstream, she arches her back and sweeps her tail several times across the gravel to create the nest site, an act she performs for the first and only time in her life. She carries just several thousand eggs and this is her only chance at reproductive success. On average, one in a thousand eggs survives to become a spawning adult.

Males arrive a few days later, prepared to compete with each other for breeding rights. Intense competition results in very few males fertilizing the eggs of many females. Once he establishes dominance over the neighboring males, he positions himself close to the female. He quivers, shudders, and crosses her repeatedly, signaling his readiness. This is the Salmon Dance, the steps choreographed over millions of years. The female releases her eggs and the male his milt simultaneously, the eggs settling into the loose gravel. With her tail she covers them with gravel and prepares to drop more into another nesting pocket within the redd. She may lay eggs in four or five spots in a single redd, following the well-worn idiom "don't put all your eggs in one basket." The practice of depositing them in several co-located pockets probably reduces the risk of losing all offspring in a single flooding event or other nest disturbance.

More work remains after laying and fertilizing the eggs. Having stopped feeding while at sea, finding their natal river through their sense of smell, forging upriver to the Buckley fish barrier and trap, hitchhiking a ride in a water tanker, re-entering the river to press on, and then laboring to locate, claim, build, defend, and fertilize a redd, the exhausted pair now guard it with their waning strength. From a few days to a month, they hold on, defending it against latecomers and upstarts. Finally, when all energy to power them is spent, the fish expire. They have closed their circle, completed their cycle of life. They have given their potential offspring a fighting chance at survival. Now their spent bodies will render one more gift.

Having acquired nearly all of their body weight while at sea, the salmon's tissues contain mostly ocean-derived nutrients. The decomposition of salmon carcasses in freshwater systems represents a huge input of carbon, nitrogen, phosphorus, and other nutrients into rivers and lakes that by comparison with ocean environments are relatively nutrient-poor ecosystems. The image of bears or eagles feeding on spawned out salmon is familiar to most people, but the nutrient transfer extends well beyond these iconic animals. Hundreds of invertebrates and 137 species of vertebrates benefit from salmon in the ecosystem, 82 of which make direct use of salmon carcasses. Those with the strongest relationship to salmon within the Puyallup River watershed include bald eagle, black bear, common merganser, harlequin duck, osprey, and northern river otter. Like the beaver and pileated woodpecker, Pacific salmon are a keystone species with a critical role in sustaining their ecosystems. Unlike those organisms, however, the nutrients in the body tissue of the far-ranging salmonids bind aquatic and terrestrial ecosystems like no other organism in the region or, for that matter, in North America. From bacteria and algae to other plants, to insects, and to large mammals, salmon leave their chemical mark nearly everywhere in an ecosystem. One study even found that an influx of salmon carcasses stimulated tree growth. From Douglas fir rootlets to crown needles, in the tissues of organisms from microscopic to mega, salmon make significant contributions wherever they live, spawn, and die.

A Mosaic in Disrepair

Long before people arrived in the Pacific Northwest, members of the Salmonidae family populated coastal rivers. Salmon and their relatives swam in rivers and lakes about 55 million years ago. Fish very similar to today's Pacific salmon appear in the fossil record beginning at about 10 million years ago. Of today's species, Chinook, coho, and rainbow trout evolved from one ancestral line about two million years ago. Sockeye descended from a second line at about the same time. Chums and pinks evolved from another group of ancestral fish at about 1.25 million years ago.

Stories about salmon are many and varied, and figure prominently in native culture. In the southern part of the Salish Sea, tribal storytellers, often elders, told stories that could be seen to explain natural phenomena. "Legend of the Humpy Salmon" tells how the adult fish weakens as it swims upriver to spawn. Others like "Spring Salmon and Steelhead Salmon" explain why different species run at different times and in different rivers. Still others give explicit instructions on behavior regarding the fish. In "Humpback Salmon," the fish will return as long as people do not ridicule it because of its hump back. Among the Puyallup people, another story cautioned against harvesting more chum salmon than needed. Should a person over-harvest, the fish would take the person's soul under the sea and kill them. Stories like these may have helped place limits on harvests to reduce the chances of overfishing.

A dam built on the Puyallup River at Electron began generating electricity in 1904. Built without a fish ladder, the dam cut migrating salmon off from the upper 30 miles of river for nearly a hundred years. In 2000, a 200-foot long concrete fish ladder finally reopened the upper river to salmon. Puyallup tribal fisheries projects lend hope toward re-establishing runs, but a major problem persists. The lack of a properly functioning structure that would allow out-migrating juveniles to move safely downriver jeopardizes their survival prospects. Consequently, conservation groups brought a lawsuit against the dam's operators for practices that kill fish protected under the Endangered Species Act.

The Army Corps of Engineers completed the watershed's second dam, Mud Mountain, on the White River in 1948. About 25 miles southeast of Tacoma, it manages flood risk rather than generate electricity. The earth-fill dam holds heavy rains and snowmelt in a 1,200-acre lake until engineers gradually release water into the river after high water events. The dam reduces flood danger for thousands of downriver homes, but with grave environmental costs. Flood control activities unnaturally alter the flow and sediment regimes in the river. The dam also limits the amount of large woody debris that accumulates downriver, reducing structural diversity and protective cover for salmonids.

Without a fish ladder, migrating salmon had no way to move upstream beyond the dam. The Army Corps needed a long-term solution, but settled instead for a short-term fix. It entered into an agreement with Pacific Coast Power Company to build a trap-and-haul operation to move fish above Mud Mountain. The idea was to corral salmon headed upriver into a holding pen, transfer them into tanker trucks, drive them above the dam, and then release them back into the river so that they could complete their migratory run. The two partners chose the White River Diversion Dam, about five miles below Mud Mountain near Buckley, as the site for the facility. Although not technically a dam, Pacific Coast had built the structure across the river in 1910 to divert water into a flume leading to Lake Tapps for the White River Hydroelectric Project. It too had been a fish barrier for decades. The trap-and-haul's design accommodated up to several thousand fish per year and during most years, it proved adequate. Two developments, though, have exposed the practice as wholly deficient and a hazard to fish survival. First, the number of fish entering the pen has increased in recent years. The influx of pink salmon has overwhelmed the tank truck hauling capacities, causing fish to remain in the pen for too long. Overcrowding and transport delays affected all species of salmonids. Tribal fisheries biologists at Puyallup and Muckleshoot estimated that up to 200,000 pinks went uncollected during their run years and either died without spawning or spawned in poor habitat downstream of the barrier.

In addition, the barrier's deteriorating condition has transformed it into a fish killer. The continuous flow of water over its top creates a false attraction for the migrating fish. They repeatedly hurl themselves against the dilapidated barrier, trying to continue upstream instead of moving along it toward the holding pen. Upon contact with the barrier, they slam against the rotting wood cribbing, abraded concrete, and exposed rebar that cause unnecessary injury and death to thousands of fish per year. Tribal fisheries staff reported that about 20 percent of a recent run of ESA-protected steelhead and Chinook sustained injuries or died from jumping onto or against the decaying structure.

After years of recurring controversy, NOAA stepped in. In 2014, it issued a Biological Opinion requiring the Army Corps of Engineers to build a new White River fish passage operation. The new facility must provide for a 95 percent "safe attraction" and increase the survival rate from 80 to 98 percent. The Corps responded with plans to demolish and replace the archaic structure with the country's largest trap-and-haul operation, capable of transferring 60,000 fish per day and 1.25 million per year. Due to come on line in December 2020, the $112 million structure, spearheaded by Senator Patty Murray and spurred forward by tribal persistence, will better manage the volume of fish and improve the passage of sediment and other material downstream.

The White River fish barrier saga epitomizes the salmon story for the Puyallup River watershed in particular and the Pacific Northwest in general. In the Puyallup basin and elsewhere, salmon suffer not just one threat, but a continuous series of challenges encountered in every phase of their life cycle. From egg to alevin, to fry, to smolt, to ocean, and home again as a breeding adult, human influence repeatedly affects one of the region's oldest and most venerated inhabitants. Some biologists see the salmon's life history as a journey through a series of connected habitats that have been degraded by logging, agriculture, mining, urban development, and other activities. Where habitat is impaired, they see a break in the chain of associated spaces, affecting each fish passing through that habitat. Many links in the Puyallup chain lie damaged or broken.

In spite of substantial effort by a wide array of partners, salmonid habitat continues to deteriorate and disappear. A growing human population demands more groundwater removal. More wells lead to low summer flows that impede upriver migrants. The nearshore and estuarine habitats that shelter juvenile salmon have shrunk by 97 percent from the original expanse of 5,900 acres that existed prior to development of Commencement Bay. Only about 7 percent of marine shoreline along the Puyallup River remains undeveloped and free of bulkheads, riprap, or other structures that impair productive salmonid habitat. Channels and levees that help reduce flood risk confine 15 percent of the known fish distribution waters in the watershed. This affects the fish in multiple ways. The levees and revetments exclude riparian habitat that attracts prey organisms crucial to the salmonid food web. A lack of riparian habitat means a loss of cover and shade that helps moderate summer water temperatures, threatening returning spawners. Without riparian zones, the waterways lack the woody debris that serves important ecological functions. Its systematic removal from the basin's waterways during much of the twentieth century—in the name of flood protection—created a deficit from which the rivers may never recover. Finally, levees limit access to other critical salmonid habitat. Pools, side channels, and off-channel habitat used at all life stages are inaccessible where levees predominate.

If we imagine the salmon runs of the Puyallup River watershed of 200 years ago, we might see them as part of a complex, detailed mosaic symbolizing the bounty of the region. Each tessera or individual tile represents an essential part of the greater whole: every nutrient cycle, ecological process, organism, species, habitat, community, and ecosystem. Together the thousands of tesserae create the andamento, or flow, that rolls from Mount Rainier's glaciers to the bay, creating a magnificent work of art.

In our short time here, we have damaged and defiled the precious mosaic. It stands diminished and so do we, made less by our longing for more. Yet, hope lives here. What we know about resilience and restoration of natural systems tells us that we can reverse the trends and repair the mosaic. Public and private agencies and organizations, municipal and tribal governments, all powered by creative and committed people, have mobilized on the mosaic's behalf. Not only the salmon's future, but also the ecological health of the entire watershed rides on their efforts—our efforts. Success or failure in the coming years will determine whether the mosaic falls into even greater disrepair or undergoes a resurgent, transformative overhaul.

Hope for the Smolt

It takes a long time to turn the Queen Mary, so goes an old saying, and the same often holds true for large-scale ecological restoration projects. If troubled by accountability issues, a patchwork approach, or insufficient funding, work can proceed more slowly than expected.

Frequently short-staffed and underfunded, federal agencies have come under fire for not fulfilling their responsibilities for salmonid protection and enhancement. The protracted debacle at the Buckley fish barrier is just one example. Others include sediment release from Mud Mountain Dam that damages habitat, levee maintenance standards that prevent riparian growth from establishing along waterways, and lenient well permitting procedures that reduce in-stream flow and impede upstream salmonid migration.

Considering that migrating salmon pass through a network of linked habitats during their lives, it follows that the most successful projects use an ecosystem or watershed approach. Projects that restore habitats and the connectivity between them have the greatest chance of long-term success. Like most river systems, the Puyallup and its tributaries flow through multiple jurisdictions. This can lead to conflicting ideas and practices about development or conservation, which hinder vision and action. As a result, isolated restoration projects can create a hodgepodge that falls short of long-term goals.

The greatest obstacle to watershed restoration regionally and within the Puyallup system is insufficient funding. A recent Puget Sound Partnership *State of the Sound Report* found annual funding shortfalls of $300 million for habitat restoration and $250 million for stormwater projects. The same holds true in the Puyallup drainage, where requests surpassed $1 billion during a three-year funding cycle but were funded at only one-third that level. Capital improvement projects like habitat creation, restoration, or setback levee construction tend to secure the most money. Lagging behind but equally important is funding for non-capital projects like public outreach and education, scientific research and monitoring, or the purchase of lands for habitat protection. During the same three-year period, non-capital projects in the basin received just 6 percent of the total $5 million request.

Despite these challenges, a handful of projects in the lower watershed offer hope for the future. The most promising projects are on the floodplains. These low-lying, flood-prone, nutrient-rich areas next to streams and rivers are ecological treasures. They provide valuable cover, resting, and feeding grounds for salmonids. The salmon in turn attract predators like birds, raccoons, river otters, and other vertebrates. Riparian woodlands along the floodplains create habitat complexity for these and other organisms. People use them too, enjoying access for fishing and other recreational activities. In other areas, the rich soils make for highly productive farmlands.

Floodplain restoration is not new in the Puyallup basin. It began in the 1990s in response to a series of 15 flood disasters within a 50-year period. The Puyallup River Basin Comprehensive Flood Control Plan emplaced goals and regulations to guide floodplain policies and projects that place it among the most progressive in the country. Moreover, Pierce County's purchase of nearly 2,000 acres of property and homes susceptible to flooding helped break the cycle of recurring flood damage. The county removes all structures and then holds the land vacant. But it was the Ford property setback levee in 1997 that first combined flood risk reduction with habitat creation and enhancement. Partners built a new levee behind the original one to give the river more room to move on its floodplain. This reduced flood hazard and reconnected over 120 acres of floodplain with the Puyallup, reopening salmonid habitat. Another project, the Soldier's Home setback levee in Orting, rejoined another 55 acres of river to floodplain. Fisheries experts believe that levee setbacks that create off-channel habitat are one of the best ways to help the recovery of Puget Sound Chinook populations.

Encouraged by the multiple benefits of flood risk reduction, floodwater storage, habitat creation, and wetlands reconnection, Pierce County prioritized a list of 32 sites for setback levees and floodplains restoration. The work received a huge boost when the 2013 Washington State Legislature allocated $50 million to the state's Department of Ecology. Joining The Nature Conservancy (TNC) and the Puget Sound Partnership as organizational leads, the trio coordinated the state's groundbreaking Floodplains by Design Initiative (FbD). They helped local groups and interested stakeholders draft and submit proposals for projects demonstrating an integrated approach to floodplain management. Twenty-four projects, 11 in the Puget Sound region, received funding in the first budget cycle.

One of the first FbD-funded projects in the Puyallup basin was the Calistoga Setback Levee near Orting. Nearly annual floods had released sediments that damaged salmon habitat. High water covered roads, suspended traffic, and created a public safety hazard. Partners that included

The award-winning Calistoga Setback Levee in Orting runs parallel to, but lies back from, the Puyallup River. It allows water to exceed the banks without flooding private property or the roadway, at the same time creating habitat for salmonids. *Parametrix, by Matt Kastberg*

TNC, the Puyallup Tribe, local governments, and state agencies worked with Parametrix, an engineering, planning, and environmental sciences firm, to move and strengthen the existing levees. They not only improved flood protection for the city of Orting, but also reconnected 42 acres of floodplain habitat to the Puyallup River. The project's innovative design and effectiveness won the City of Orting multiple achievement and recognition awards. Better still, the new levee passed its first test, keeping roadways open when a major storm brought high water.

More work delivered similar results. One project sought to benefit both agricultural and restoration interests; another entailed land acquisition. Encouraged by the successes, Pierce County created a 10-year plan for floodplain reconnection to build nine more setback levees for $200 million. Competition for FbD funds rose sharply in the next funding cycle, however, and to make matters worse, program funding suffered a 30 percent cut. Work in the Puyallup basin would proceed in much smaller steps toward the grand vision. Funding of one FbD multi-year, $9.2 million model project will reduce flood risk, acquire habitat, and create, protect, and rejoin hundreds of floodplain acres with tidal marsh habitat. Given adequate funding, Floodplains by Design could become a national model for collaborative floodplains management.

Similar initiatives without FbD funding are also underway in the lower watershed. King County spearheaded an $18 million effort that reduced flood risk to over 200 homes in Pacific and Sumner and reconnected 121 acres of off-channel habitat with the White River. Planning continues for a nearby companion project, also led by King County.

South Prairie Creek is the most important salmonid producing waterway in the Puyallup River watershed, home to ESA-protected Chinook and steelhead. Its floodplain restoration is a partnership between Pierce County, the Pierce Conservation District, the South Puget Sound Salmon Enhancement Group, and the Puyallup Tribe. The project, at an old dairy farm, will place large woody debris in a half-mile stretch of creek, demolish vacant buildings, and cover 18 acres with native plants. The sum of these undertakings holds great promise for the watershed—and the trend region-wide gives cause for optimism. The Puget Sound Partnership's 2015 *State of the Sound Report* found that 39 floodplain projects restored 14,500 acres of habitat in a four-year period. Estuaries recovered another 2,260 acres of tidal processes. Other data indicated a slowing down of shoreline armoring projects and the removal of some existing structures. In Puget Sound, according to the Partnership, habitat restoration efforts "are making progress, reflecting investments over time in restoring estuaries, floodplains, and riparian corridors to benefit salmon and other fish and wildlife habitat."

Not long ago, a friend joined me for several drab winter days exploring the waterways, tideflats, and restored sites in the lower Puyallup. A 15-acre site named Yowkwala, meaning "eagle" in Lushootseed, nestles on Commencement Bay's east side. A high-pitched chittering grabbed our attention and looking up, we spotted four bald eagles circling overhead. They played follow-the-leader and did barrel rolls, taking advantage of the afternoon onshore breeze. The perfect light of a sunbreak glinted off their regal white heads and tails, and sturdy, bright yellow feet and bills, reminding us of the importance of the work.

GREAT MOUNTAIN, GRAVE DANGER

Imagine the surprise felt by construction workers at a new housing development in Orting in 1993 when, while excavating for a new sewer line, they uncovered the lower portion of a giant Douglas fir, its base rooted 16 feet underground. Not yet petrified, it was part of a vast buried forest. Along with dozens of other trees, including some that may have floated in, the workers had inadvertently dug up priceless clues about one of the largest and most recent lahars to rumble down Mount Rainier's slopes.

Because of his keen interest in buried and submerged forests, geologist Patrick Pringle set about unlocking the secrets held in the Orting tree rings. With colleague Jim Vallance, he used a process known as carbon-14 wiggle matching to sample one tree that was more than 360 years old and another a hundred years older. Their results dated the Electron Mudflow to about 1500.

Homes and other structures built in Orting in the early 1990s—and the entire city for that matter—sit atop this most recent great lahar to rush down the Puyallup River valley. A sector collapse of the mountain's upper west flank near Sunset Amphitheater triggered the flow that traveled at speeds upwards of 40 miles per hour and, at Electron, was higher than a 10-story building. Within an hour, thirty feet thick, it surged through Orting, leaving behind a 16-foot layer of concrete-like slurry en route to its runout near Puyallup. A gargantuan boulder 16 feet across rests on a lawn across from one of Orting's schools, a reminder of the flow's herculean ability to transport material. Geologists estimated the flow's volume at between 260 and 340 million cubic yards.

A Native American story appears to explain this incredible force of nature. In "The Young Man's Ascent of Mount Rainier," the main character climbed in search of his spirit power. Once on top, he bathed and swam in an underground lake. The mountain told him, "At the time of your death, when you are very old, my head will burst open and the water which you see now [in the lake] shall flow down the hillsides." According to the tale, when the man died, water rushed down the mountainside and swept away all of the trees. The local Indians called the place, which is present-day Orting, *swe'kW*, meaning "an opening or clearing in the forest."

The antecedent of the Electron Mudflow began many thousands of years ago amid a period of hydrothermal alteration. During an eruption, buoyant molten rock rose toward the surface and entered cracks in the volcano. As these dikes cooled, they emitted gas and heat into the groundwater. At temperatures ranging between 300° and 600° Fahrenheit, the broiling, acidic waters began transforming the hard volcanic rock. Breccia, a permeable and porous sedimentary rock made up of pieces of older rocks, is especially susceptible to these conditions. The acidic water and heat gradually converted the breccia into a water-saturated, soft and slippery, mineral-laden, clay-rich rock. These changes compromised its strength. Eventually the formation could no longer support its own weight and the weakened mass avalanched downslope.

Hydrothermally altered rock occurs on other Cascade volcanoes besides Mount Rainier. Mount Adams and Mount Baker possess large volumes of breccia-altered rock that may someday give way to a massive sector collapse and lahar. Some geologists believe that these volcanoes' similar eruptive styles, characterized by abundant lava flows and breccias high on the mountain, produce the perfect conditions for hydrothermal alteration. This contrasts with other Cascade volcanoes that tend to form lava domes instead of generating flows. Mount St. Helens, Oregon's Mount Hood, and California's Mount Shasta and Lassen Peak all have low volumes of breccia in their upper regions and have little hydrothermally altered rock. Experts consider none of them at high risk for lahars.

Although there are practically no hydrothermally active areas on Mount Rainier today, enormous deposits of weakened rock remain on the mountain's upper slopes. The Puyallup River valley holds the greatest volume, followed by the Nisqually River drainage. Geologists use three intensity levels to describe the degree of alteration. They found that rock in Sunset Amphitheater and at the head of the Puyallup Cleaver was in the "most strongly altered" category. Hydrothermally altered rock shows little magnetism, so one groundbreaking study used a low-flying helicopter to determine the magnetic and electromagnetic properties of rocks in the area. The results allowed researchers to map and estimate the volume and extent of the fragile, rotten rock. The results were terrifying. Approximately two billion cubic yards of hydrothermally altered material sits above the Puyallup basin, a volume six to eight times greater than the Electron Mudflow. When the mass releases—and there is no reliable way to predict its timing or whether it will be several events—it could unleash a catastrophic lahar down the valley surpassed in size only by the Osceola Mudflow.

Nowhere else in this study of Mount Rainier do we encounter such a potentially calamitous collision between the place and its people. The Puyallup River watershed, one of Washington's most populous, lies directly down valley from a rock mass that may become a devastating lahar, possibly the mountain's largest in the last 5,600 years. If the hydrothermally altered rock represents "the unstoppable force" once it breaks loose, and the urbanized lowlands with its thousands of people "the immovable object," how do we prepare?

ANTICIPATING THE UNTHINKABLE

It is 8 a.m. in the briefing room for Orting's annual lahar evacuation drill, and after a cursory welcome the police chief launches into an orientation with maps, procedures, and radio instructions. The nearly 40 law enforcement officers and emergency responders listen intently, taking notes. This is no coffee and donuts affair—these folks are all business. Their goal is to oversee the safe passage of 2,000 students and community members to "the safety zone," a spot on the road 30 feet above the valley floor, within 45 minutes. The mayor exhorts us, saying that, "Just yesterday, my wife pushed her mother in her wheelchair to the safety zone in 28 minutes." He fails to mention that it's nearly two miles from the elementary school to the safety zone, and that kindergartners are notorious for having short legs. Police and fire personnel review the incident action plan, verify that we know our roles, responsibilities, and duty stations, and then send us into the field.

Despite some confusion about whether to stay on the sidewalks or surge onto the road, the drill proceeds beautifully. I watch hundreds of students and teachers stream past, a human river seeking the safety of higher ground. Dozens of staff and volunteers in safety vests direct traffic and encourage the students. "Keep going! You're looking great!"

Remarkably, 93 percent of the drill's participants made the safety zone within the allotted time, a record in comparison to previous drills. As I walked back to town for the debriefing, I stopped to look upriver. The tranquil, emerald green watercourse with its few logs and downed trees belied the reason for the drill and the possibility that such a peaceful scene could transform into a roiling cauldron of chaos.

Before the advent of coordinated evacuation drills, Orting parents and teachers became alarmed for their children's safety when they learned of the hazard and in 2001 formed Bridge for Kids, a grassroots organization dedicated to finding a fast route to safety. The group's vision calls for two pedestrian bridges designed to evacuate up to 12,000 people within 40 minutes. The snag is in the funding for a bridge across the Carbon River to a path ascending the valley wall to a safe haven. At an estimated cost of $42 million, it will be the most expensive pedestrian bridge in the country. Bridge for Kids also reaches into the community, helping residents understand the impending danger and teaching them how to prepare for a lahar. They invite the public to join in monthly evacuation drills, improving the prospects for a successful evacuation.

The Washington Soldiers Home is another example of lahar preparedness in Orting. Operated by the Washington State Department of Veterans Affairs, the residential facility serves U.S. Armed Forces veterans. The staff oversees evacuation exercises that prepare for earthquakes, fires, and lahars. As partners with the Pierce County Department of Emergency Management (PCDEM) and other agencies, they believe that more than money, time and preparation make for the best defense against an emergency.

Since white settlers began moving into the region in the 1800s, the population in the six lowland valleys that drain the mountain has swollen to more than 2.5 million people. Over 90,000 live within an area that geologists call the Mount Rainier lahar hazard zone. One study estimated that a Puyallup valley lahar similar to the Electron Mudflow could exceed $12 billion in damage to property, buildings, and their contents. Based on the dozens of events in the last 10,000 years, geologists estimate approximately a 10 percent chance of one reaching an inhabited area over the course of an average lifetime.

The highest stakes and greatest danger lie in the down valley, lowland areas of Orting, Sumner, Puyallup, Pacific, and Fife. Orting, Sumner, Fife, and Carbonado (which lies in the Carbon River drainage) have the highest percentages of residents and assets within the lahar hazard zone. All of the K–12 schools in Orting, Fife, and Carbonado are at risk of lahar inundation; half of Sumner's schools sit on the valley floor. Of the 27 communities within the zone, Puyallup has the most people (nearly 20,000), the most assets, and the most businesses at risk. Analysis of a composite index based on multiple factors determined that Puyallup, Sumner, and Fife have the highest combinations of numbers and percentages of people and assets at risk to lahars.

The 1980 Mount St. Helens eruption gave researchers a better understanding of Mount Rainier's volcanic tendencies. In a similar way, lahars around the world help raise awareness and shape our preparedness. Over the last hundred years worldwide, lahars have caused over 30,000 deaths, left 93,000 people homeless, and disrupted the lives of another million people. Hundreds died in the Mount Pinatubo, Philippines, eruption and ensuing lahars. Heavy rains in northwestern Nicaragua in 1998 triggered a lahar that buried two towns and killed more than 2,500 people. The 1985 Nevado del Ruiz eruption was the most catastrophic lahar in recorded history. Over 23,000 perished in the city of Armero, a hundred miles northwest of the Colombian capital of Bogotá, when overrun by a lahar following an eruption. Another 50,000 people were left homeless or otherwise affected. A subsequent study showed that a poorly informed public and inadequate coordination between municipal officials, scientists, and emergency management personnel contributed to the disaster's scale. The Armero catastrophe sent a startling wakeup call around the world. To share information and reduce the consequences of a Puyallup valley lahar, PCDEM representatives traveled there to meet with local officials. The trip created an exchange program between Colombia, Ecuador, Chile, and Washington that now allows officials to collaborate on preparedness.

Scientists and public officials have developed four strategies to reduce and manage lahar risk: hazard avoidance, modification, warning, and response/recovery. In terms of hazard avoidance, some people or businesses may relocate to avoid lahar danger, but the wholesale migration to higher ground is an unlikely and impractical solution for Mount Rainier's lowland communities. Hazard modification is also unlikely, given the scale of the impending lahar. A series of check dams, slit dams, and other collection devices known as "Sabo works" was developed in Japan in the seventeenth and eighteenth centuries, leading to similar projects in Taiwan, South America, and Europe. No plans are underway for hazard modification on the Puyallup.

In the Puget Sound region, hazard warning and response/recovery form the foundation of lahar hazard management. The backbone of Mount Rainier preparedness is the lahar warning system, which can alert thousands of people in the valley of an approaching flow. It does not eliminate risk, but if properly executed, it is the area's best chance of reducing it. The United States Geological Survey (USGS) and the PCDEM developed and brought the system into test mode in 1998. Recent upgrades have revitalized the aging system.

The Mount Rainier plan has three integrated components. First, an automated network of seismic sensors called acoustic flow monitors "listen" for an approaching lahar. The half dozen sensors in each of the Carbon and Puyallup River upper valleys stand guard as the first line of defense. A geophone embedded in the ground records vibrations within a frequency of 30 to 80 hertz (low E on a bass guitar registers at 40 hertz). A circuit board analyzes the incoming data and transmits it down valley for a closer look. Tripwires on the monitors also trigger the alarm.

Second, the USGS, PCDEM, Washington State Emergency Operations Center, and South Sound 911, the area's inter-local emergency dispatch center, all evaluate the information obtained from the flow monitors. In the event of a lahar, the operations center and 911 verify their conclusions with each other before the system automatically alerts emergency management and law enforcement agencies.

Third, a notification system alerts people to its approach, using as many means as possible to notify citizens of the imminent crisis. The 33 outdoor warning sirens lining the valley floor from Orting to the Port of Tacoma will sound off. The Emergency Alert System (also known as the national public warning system), NOAA weather radios, radio, and television will broadcast prepared warnings. Pierce County Alert, a reverse 911 system, will notify subscribers.

The hazard response/recovery component depends on two critical factors. First is the integrated warning system that prompts the evacuation of people in Puyallup basin lowland areas to higher ground. Most local evacuation plans call for people to get at least 30 feet above the valley floor. Since lahars adhere to the principles of gravity and generally flow down river valleys, people already above the valley floor will be out of harm's way. Second, federal, tribal, state, county, and local staffs must gather regularly to review their responsibilities and planned responses. This is essential, given that multiple jurisdictions within the hazard zone can complicate the response and recovery effort.

The USGS estimates of lahar travel times amplify the importance of the warning system. Orting, the community nearest the mountain, lies about 32 miles from the likely point of origin. Current estimates predict that a lahar could reach Orting within 42 minutes after sounding the alarm, and Sumner and Puyallup in more than an hour. The runout to Tacoma's Commencement Bay will take close to two hours. Experts caution that after such an event sedimentation and flooding may occur for decades or centuries, preventing residents from returning to lahar-affected areas.

In spite of the dire prospects surrounding a potential Puyallup valley lahar, continued research and improving technologies will provide more advance notice and evacuation time. Broadband seismometers will gradually replace the aging acoustic flow monitors, sending more information more quickly. New software will help make full use of the data, too.

Also important is the Pierce County Department of Emergency Management's continued advocacy for public awareness and preparation. The department's Volcano Evacuation Route roadway signs direct motorists to higher ground. The PCDEM tests the warning system sirens at noon on the first Monday of each month. More importantly, they help valley residents understand the hazards and take all necessary steps to prepare for a lahar. This includes knowing and practicing evacuation routes and having a family emergency plan and kits.

Curious, I drove to Orting one spring morning to observe a siren test. Promptly at high noon, the wailing began, carrying on for an ear-piercing five minutes while life carried on all around me. It seemed odd that the locals continued their midday errands and that the elementary school students continued chasing each other during their recess, until I remembered—and was reassured by—the lahar preparedness workshops and evacuation drills, the sensors and tripwires lying in wait on the mountain's flanks, and the vigilance of scientists, emergency personnel, and municipal officials.

6

The Carbon River Area: Land of Moisture

Mount Rainier from the terminus of the Carbon Glacier. The river's headwaters rush down valley from the cave-like opening at the glacier's toe. The U-shaped valley results from centuries of glacial action. *Drawing by Lucia Harrison*

DAMP VALLEY, LUSH FOREST

The lowering clouds timed it perfectly. I no sooner stepped out of the car at the Carbon River entrance than it began to rain. I abhor the necessary evil of restrictive, clammy raingear, and since this would be a short hike, I left it to my layers of poly and wool to absorb the rain yet keep me reasonably warm.

157

This was an early spring trip, its purpose to observe and welcome the buds, shoots, and other new growth in this lush old growth forest on the mountain's northwest side. Except for midsummer's ephemeral drought season, rain is a constant companion here. On the rare occasion it doesn't rain, something obvious is missing that I can't quite put my finger on, like I've left my lunch at home but haven't realized it yet. Rain is the defining characteristic of the Carbon River area, the thread stitching the valley to itself, the answer to the questions, "Why is it so green here?" "How come these trees are so huge?" "Why do mosses cover everything?" The answers are rooted in rainfall, anchored in the annual precipitation rate. A strange beauty inhabits the dense, ethereal mist, the seemingly endless drizzle, the occasional pelting rain. It is no wonder that these forests evoke reverence, poetry, and art. Moving beyond my own discomfort, I grasp the true Carbon River rain spirit. A Mount Rainier rite of passage, it is an essential experience to rain hike this valley; the newly embroidered patch for your daypack, a merit badge for your sash. If the weather is clement on your next trip here, make immediate plans to return at another time when it's sure to be raining.

The first stop in today's drizzle is a forest opening that hosts a thick stand of the spiny and sprawling devil's club. Knock-kneed and crooked, this abundant member of the forest's luxuriant understory shows off fragrant new leaf buds atop stems ranging from three to ten feet tall. Its scientific name, *Oplopanax horridum*, hits the mark. *Oplo* is Greek for armor, referring to the protection afforded by its thorns. It combines with *panax*, meaning cure-all, to reflect its wide-ranging medicinal properties. *Horridum* becomes obvious the moment that one encounters the fearsome spines that grow up to a half-inch long.

The deciduous canes will soon support large, alternate, maple-like leaves covered with tiny spines, which prevent heavy browsing by deer and elk, though I often find partially eaten young leaves in May or June. Later in the season, the pyramid-shaped clusters of tiny white flowers appear at the top of the plant and then transform into a mass of shiny red berries, each smaller than a pea. Humans don't eat the berries, but bears do.

Despite its dangerous disposition, devil's club holds an important place among plants in the Pacific Northwest. Coast Salish people steeped the roots and drank the tea to treat colds. Those suffering from arthritis or rheumatism drank a tea made from the roots and the greenish inner bark. Devil's club contains antibiotics that fight a tuberculosis-causing Mycobacterium, so some indigenous people mixed the inner bark extract with other plants to brew a treatment for tuberculosis. Others believe it to be a hypoglycemic, capable of lowering the blood sugar. They drink tea made from the inner bark as a treatment for diabetes.

Aside from its medicinal value, devil's club was useful to native people in other ways. Men carved fishing lures from the light-colored, buoyant wood. The dried and pulverized bark made an effective deodorant or baby talc. And it was highly valued as a protective agent. The first people of the Pacific Northwest believed that devil's club protected them from all manner of evil influence in the same way that indigenous people in other parts of western North America believed that thorny or prickly plants afforded protection. Some carried a devil's club talisman to ward off evil. Ceremonial dancers gained protection from evil spirits by wearing face paint made from its charcoal. Shamans used it in ceremonies to acquire supernatural powers. From a medicinal and cultural standpoint, devil's club has been as valuable as it is fierce looking.

The drizzle becomes steady rain as I move up the trail. Spring's harbinger, trillium, greets me along the way. People often call this early bloomer "wake robin" where its emergence coincides with the return of the robins. Its three leaves, petals, sepals, and stigma make for a lovely, pleasing symmetry. The creamy, white flower turns pink and then purple, the color helping to attract pollinators.

Trillium disperses its seeds in an unusual way. A fleshy lobe attached to each seed contains fats and protein. Ants carry the seeds home to eat or feed the oily lobes to their larvae. They don't bother eating the seeds, which they dispose of in refuse piles, where they may later germinate and grow into the next generation of trillium. Seed-transporting ants are more common in the eastern United States, but trillium, wild ginger, bleeding heart, and inside-out flower number among the "ant plants" at Mount Rainier.

Green dominates the Carbon River color palette, with even the untrained eye able to note a nearly endless variety of verdant shades. There are over 250 green hues in ornithologist Robert Ridgway's 1912 publication *Color Standards and Color Nomenclature*, his labor of more than 20 years that sought to record all of the colors found in nature—1,115 of them. The book includes such Carbon classics as Corydalis Green, Hellebore Green, Lichen Green, and the Pacific Northwest's trademark, Emerald Green. Greens abound in several species of ferns unfurling their fiddleheads on this damp spring day. Each frond emerges tightly curled, gradually unrolling to its full length. The tip suggests the scroll at the end of a violin's neck. Native Americans ate the plentiful fiddleheads of some species; many people enjoy them today. Licorice fern, growing abundantly on big leaf maples in riparian areas, was a potent medicine. Its rhizomes—the root-like structures that propagate the plant—were roasted or baked, and used to treat coughs and influenza. The widely found bracken fern, easily identified by its three large, triangular leaf blades, stands upright and erect. Many native groups roasted, peeled, and ate its rhizomes. Bracken and the ever-present sword fern, layered in pit ovens, prevented hot coals from burning slow cooking food. Sword ferns also covered berry-drying racks and served as flooring, bedding, and medicine.

On another rainy day in a month or two, I'll come back to snack on the large and tasty salmonberry. Coast Salish people gathered and ate the berries and sprouts of this common forest shrub. Its raspberry-like fruits are too juicy to dry and keep for later, so people ate them fresh, and with salmon. Some believe this to be the source of its common name. The berries ripen about the time that Swainson's thrush, one of Mount Rainier's sweetest-

Nurse logs mark the junction of decay and regeneration in old growth forests, and are biodiversity hot spots. *Illustration by Kirsten Wahlquist*

voiced songbirds, returns from its wintering grounds between Mexico and Argentina. The bird's flutelike song spirals up as spring and summer days wind down. Probably because of its arrival with the ripening fruit, in many native languages this thrush's name translates to "salmonberry bird."

The other target on my next trip will be the closely related thimbleberry. Like its cousin, it requires both moisture and sun, preferring slightly drier locations, and people enjoyed the berries and sprouts fresh. Large leaves similar to big leaf maple often conceal the nearly round, deep red fruit. The size and shape of thimbles, the berries are often so delicate that they fall apart in your hand. Not everyone favors their nutty, gritty taste. They don't keep well until later; simply try a few where you find them.

The rain will be warmer on a trip in June or July. It will also be prime time for the eye-catching red columbine, the park's loveliest member of the buttercup family. A break in the showers allows its comely red and yellow flowers to nod sleepily in the gentle breeze. Hummingbirds and butterflies probe the blossoms for nectar. Bees try the back door, sipping nectar after clipping the spur-tips, just as native children once did.

Light comes at a premium to the forest floor, where plants use any available means to gain a competitive advantage. Devil's club, salal, and huckleberry shrubs get a boost from downed logs, growing upon them to get an edge in the battle for precious sunlight. Douglas fir and western hemlock seedlings crowd onto the nurse logs too, jockeying for position upon these old growth forest mainstays. Coarse woody debris lies everywhere, the downed logs strewn about like spillikins. The logs are probably a remnant of the devastating 2006 flood or another recent disturbance event. Snags of all sizes point skyward. Red huckleberry shrubs sprout from some of their broken tops like wild green wigs.

The riveting feature of the forest here, like other old growth stands at Mount Rainier, is the number and enormity of the trees. Western red cedar and western hemlock dominate this stand, the shade tolerant hemlock having waited to fill canopy openings. In spring and early summer, the hemlock's one to three inches of new, yellow-green growth at the end of each spray contrasts sharply with the dark green growth of previous seasons. Observant hikers can find the occasional Pacific yew—gangly, asymmetrical, and easily overlooked. Sitka spruce, rare in other parts of the park, flourishes here. Behemoth Douglas firs are common, the deeply furrowed bark protecting them against fire. In the absence of disturbances, these statuesque old-timers can live to over 500 years.

The rain that supports this lush old growth forest begins thousands of miles west, far out over the Pacific Ocean where the region's weather originates. Semi-permanent high- and low-pressure areas over the North Pacific move air eastward over temperate waters. The air absorbs moisture while gradually warming to nearly 50° Fahrenheit, even in winter. The jet stream serves as the conduit, a massive river of air currents that flows hundreds of miles wide and several miles deep through the stratosphere. After days or weeks the moisture-laden mass, now in prime cloud-making condition, collides with the mountains of western Washington. The Olympics and the Cascades disrupt and redirect airflow that shapes the region's weather and climate. The Cascades, for example, moderate winter temperatures by keeping the icy blasts of the continental interior from entering western Washington from the east.

Warm, moist air from the Pacific, the high and low pressure cells that direct it, and the mountains it encounters combine to create the temperate maritime climate of the Puget Sound lowlands. This is a Mediterranean climate, named for southern Europe's similar climatic pattern. During the wet and cloudy winters spanning from November to February, weather typically moves from the southwest to the northeast. A long, spring-like stretch extends from February to July, setting up the warm and dry, brief summer that reaches into September. From then until November the weather transitions back to rainy but mild conditions. Annual precipitation in the Seattle area averages around 38 inches, less than the yearly rainfall in Boston, Miami, or New York. The number of cloudy days per year in the Puget Sound region far exceeds those in each of the three cities.

The effects of these basic meteorological principles of air movement, cloud formation, and the resulting precipitation greatly intensify in Mount Rainier's Carbon River basin. Winds approach the mountain from the west and northwest during the winter. The valley's northwest orientation makes it a perfect catchment basin for the moist air moving in from the Pacific Ocean. This combination of prevailing wind direction and the lay of the land profoundly influence the local weather and climate. Moist air parcels move up valley, forced upward over the terrain. In this process known as orographic lifting, the air expands, cools, and moistens as it flows to higher elevations. When fully saturated, clouds form. Eventually the moisture load becomes too heavy for the air parcel to hold it any longer and precipitation occurs. Rain falls in the lower reaches and snow further up as the clouds move up and over the mountain. They drop most of their moisture

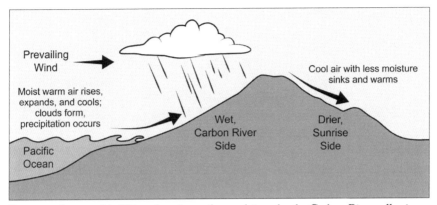

Diagram of orographic lifting, the science that explains why the Carbon River valley is wet, green, and breathtaking. *Kirsten Wahlquist*

on this side, leaving the eastern slope areas of Sunrise considerably drier. Precipitation averages 70 inches per year here, nearly double the average in Seattle. Rain and snow help create the Carbon's old growth forest and rich, emerald understory.

WONDERS LARGE AND SMALL

Prodigious trees and year-round greenery may be the valley's most conspicuous attributes, but its temperate, moist climate gives rise to other unusual features. Its northwesterly orientation not only funnels moisture up the mountain, but it also limits solar input. The western exposure causes shade to linger for most of the day during even the short but sunny summer, contributing to low snowpack melt. More snow remains longer and at lower elevations than anywhere else at Mount Rainier, making the Carbon Glacier a standout in several ways. With an area of 2.8 square miles, it's the mountain's third largest behind the Emmons and the Winthrop. Its 700-foot depth at mid-glacier makes it the thickest. And its volume of one billion-plus cubic yards of ice and snow give it the greatest volume of Tahoma's glaciers. Finally, as the park's longest glacier, the Carbon's icy river flows 5.2 miles before terminating at 3,550 feet in elevation, lying lower than any others in the contiguous United States. The 17-mile round-trip hike to see its snout makes for a long day. I once overheard a visitor remark, "This glacier is kind of boring," but few people have one within driving distance of their home.

The moss-covered limbs, downed logs, and lush understory characteristic of the Carbon River rain forest. © *Photo by Stephen Penland*

Mount Rainier's height and girth, the valley's orientation, and subsequent climatic patterns make the Carbon River valley a temperate rain forest similar to those on the Washington coast. The Carbon River Rain Forest Nature Trail, a .3-mile loop that begins at the gated entrance, is an excellent way to explore and ponder the forest's photosynthetic marvels. Like the renowned Hoh and Quinault rain forests at Olympic National Park, draping mosses and crumbling nurse logs mark the intersection of decay and renewal. In imperceptibly slow motion, logs from toppled trees break down and rich new soils form. The forest dies and regenerates simultaneously.

We humans often quibble, and forest ecologists are no exception, disagreeing whether the Carbon is a "true" temperate rain forest. It exceeds the general guideline of at least 55 inches of annual precipitation, but its average summer temperature of 68° Fahrenheit is outside the average of 50° that is used to describe a rain forest. Aside from its temperature range, however, it matches up closely with the Olympic rain forests. Trees there range taller and mossier in terms of diversity and abundance than the Carbon's trees, but some here rival those on the coast. A canopy researcher who worked at both Mount Rainier and Olympic National Parks found the forests strikingly similar. The bickering may continue, but some of us will call it as we see it: a beautiful, lone example of an inland temperate rain forest.

Walking the road toward Ipsut Creek, I find a proliferation of plants growing upon other plants. Cat-tail moss cloaks countless tree limbs. Red huckleberries stake their claim to the side of a live Douglas fir. Vanilla leaf, deer fern, foamflower, and several types of moss crowd the base of another. These plants number among the nearly 30,000 species worldwide that can grow as epiphytes. From the Greek word *epi* meaning "upon" and *phyto* meaning "plant," they grow directly upon other plants without the benefit of soil. They are not parasitic, though. Causing no harm to their hosts, they receive only physical support from them. They obtain their nutrients from elements dissolved in the abundant rain and mist. Known as commensalism, the relationship benefits the epiphytes without affecting the host plant.

Epiphytes are widespread and in our region are common in managed, old growth, and rain forests. Their diversity peaks in forests older than 200 years. Logging on 40- to 80-year cycles impacts biodiversity, making protected areas like the Carbon and those in the Olympics critically important. Colorful pollinator-attracting vascular epiphytes flourish in tropical rain forests, but not in the Pacific Northwest. Except for wall-to-wall licorice fern clinging to tree trunks and some forest floor examples, most Northwest epiphytes lack the vascular tissue that transports water and nutrients. These are the non-vascular plants, the bryophytes, which include mosses and liverworts.

Mosses are those tiny, nondescript plants that eke out entire, little-noticed life cycles in a wide range of environments, from urban sidewalk cracks to the edges of the wildest rivers. They begin as single-celled spores, the simplest of all land plants. If the spore lands on a moist substrate, it germinates and transforms into a gametophyte, the noticeable part of the plant. The gametophyte then develops male and female reproductive organs. Sperm swim through the available surface water to the eggs to fertilize them. When a capsule on the gametophyte releases the spores, the cycle of life repeats itself.

Despite lacking the roots, flowers, seeds, and water/food transport systems of flowering plants and their kin, 22,000 species of mosses grow worldwide. About 900 live in northwestern North America with more than 130 species of epiphytic mosses inhabiting Pacific Northwest rain forests. Besides cat-tail, common mosses here include Oregon beaked, step, lanky, and the several difficult-to-distinguish species of *Dicranum*. A handful of any of these is likely to hold thousands of invertebrates that range from protozoa to mites to springtails. Mosses and their inhabitants are another microcosm of the biodiversity of old growth forests.

Our other major epiphyte group is the lichens, an ancient and prolific group of organisms. About 1,200 species occur in the region; one study found nearly a hundred in a southwestern Washington old growth forest. Lichens are composite organisms consisting of a fungus and either or both an alga and blue-green algae, known as cyanobacteria, which first appeared in the fossil record over three billion years ago, supplying oxygen to the atmosphere. Two billion years later, the earliest land-based forms began stabilizing soils. Together, the appearance and function of these fungal-algal associations differ from their free-living forms, their symbiotic relationship allowing them to live where none of the partners could survive individually. Lichens lack roots and obtain most of their moisture and nutrients from the atmosphere or substrate to which they attach. This tends to make them slow growers: a four-inch diameter lichen may be more than 25 years old. Their use as food, shelter, and nesting material by wildlife give them an ecologically vital role in the region's forests. Moreover, their sensitivity to environmental pollutants makes some lichens reliable indicators of air quality.

Here's how lichens work: The fungus gives the organism its form and most of its biomass, including an inner layer that helps absorb and store water and nutrients. It provides structural support and a stable microenvironment for the alga to live. In return, the single layer of algal cells uses carbon dioxide and water, via photosynthesis, to produce carbohydrates that nourish the organism.

Two algae and a fungus combine to form lettuce lung (*Lobaria oregana*), one of the most common yet valuable lichens found in the northwest and at Mount Rainier National Park. Hand-sized and leaf-like, its yellowish-green color helps identify it. Under ideal conditions, its biomass may total more than 800 pounds per acre. One outstanding characteristic is its ability to absorb nitrogen from the air and convert it to a usable chemical form. Nitrogen is often scarce in old growth forests, so this nitrogen-fixing ability, as biologists call it, is a much-needed ecosystem service. The nitrogen later becomes available to other organisms through litter fall or leaching.

Since mosses and lichens lack the vascular systems that other plants use to acquire forest floor nutrients, they compensate in other ways. First, epiphytes often form large mats that serve as catch basins for rain, mist, or fog and the nutrients they contain. Second, the mat's chemical composition helps it trap and retain nutrients. Negatively charged sections retain cations—positively charged ions—that would otherwise quickly pass through the mat. Third, drought tolerance allows some epiphytic mosses and lichens to suspend metabolic processes and stop photosynthesizing during times of extreme heat. This keeps them alive during the Carbon River's hot spells that kill some vascular plants. They resume water absorption and photosynthesis as soon as the rain or mist returns.

Epiphyte communities locate themselves along a vertical gradient in old growth forests. Mosses and liverworts dominate from ground level up to the middle of the light transition zone, about 80 feet off the ground. They gradually give way to lichens containing cyanobacteria (called the cyanolichens), which concentrate between about 43 and 120 feet above the forest floor. Lettuce lung lives in this zone. Pendulous, hair-like lichens called the alectorioids replace them in the higher zones of the forest canopy. Common species include *Alectoria* and *Bryoria*, favored foods of northern flying squirrels, and *Usnea*. Their combined biomass is substantial, totaling over 1,100 pounds per acre.

Competition for sunlight in old growth forests is so intense that the higher an epiphyte can grow in the canopy, the better its chances to obtain the radiation needed for photosynthesis. Free from competition with the mosses and ground-dwelling vascular plants, they intercept water and airborne nutrients that enable them to thrive. The forest canopy, once called "the last biotic frontier," was largely unexplored until the 1970s when dauntless forest ecologists began using climbing equipment to access it. First in Oregon's old growth forests, then in Costa Rica, and then on Washington's Olympic Peninsula, and later around the globe, researchers began making strides toward understanding this aboveground world.

The forest canopy includes the crowns of all trees in a stand, its crown vegetation, its epiphytes, and the air spaces between them. These component parts create structural complexity and make it an integral part of the forest ecosystem. Epiphyte biomass can exceed 2,100 pounds per acre—about the weight of a small car. This abundance increases its surface area and biodiversity by providing niches for canopy-dwelling organisms. An untold number of microorganisms, plants, insects, birds, and mammals inhabit this space. Some of them rarely venture onto the forest floor, if ever. Epiphytes also contribute carbon and nitrogen to the forest. In one strategy, precipitation leaches nutrients from canopy epiphyte mats and transports them directly to the forest floor or through stem flow. In another, wind, wildlife, and falling trees or branches carry them earthward, where other organisms absorb their nutrients.

A couple of discoveries surprised the first ecologists using climbing ropes, mechanical ascenders, and seat harnesses to explore forest canopies. Over hundreds of years, the epiphytes that live, die, and decompose on branches and in branch junctions high overhead form a mat of arboreal soil very similar to that on the forest floor. Known as canopy soils, they create habitat and hold food and water for the next generation of epiphytes and other organisms. Their thickness on big leaf maple branches in the Olympic rain forest ranges from five to nineteen inches. Sitka spruce canopy soils measure between four and seven inches thick.

Researchers wondered whether there were roots growing in the forest canopy soils, and sure enough, Nalini Nadkarni, then a graduate student, found them growing under a thick moss mat in the Olympic rain forest. She traced them back to their origin at a branch junction to determine that the roots came from the host tree. They ranged from tender, young rootlets to well-developed ones four inches in diameter, identical to the subterranean roots of the forest floor. She later found canopy roots in tropical rain forests in Costa Rica, Papua New Guinea, and New Zealand, proving that they were a natural response to the availability of water and nutrients held in the mats and soils. Canopy roots create a significant shortcut for the host tree, enabling it to access precious resources directly from the epiphytes without having to wait until they accumulate on the forest floor through litter fall.

Arboreal roots have other characteristics similar to their belowground counterparts. Big leaf maple canopy roots have mycorrhizae just like those beneath the forest floor. In this symbiotic relationship between the root and a fungus, thread-like filaments called hyphae attach to the roots, increasing their ability to absorb nutrients from the soil. Another symbiotic mechanism, root nodules, also mimics underground processes to help canopy roots obtain nutrients. The nodules contain rhizobium bacteria that convert atmospheric nitrogen into a useable form for the host tree.

Forest ecologists and climate scientists remain uncertain about the effects of climate change on epiphyte communities and the forests that host them. Most regional models project a decreasing snowpack accumulation—but an increase in winter rainfall—and a drop in summer precipitation as temperatures continue rising this century. Canopy communities are sensitive to environmental changes because they inhabit the intersection of terrestrial and atmospheric worlds. Ecologists project changes will come to the Puget Sound region's forests because of increasing air temperatures, snowpack reduction, and reduced summer rainfall. Douglas fir habitat may shrink in the lower reaches of southern Puget Sound and in the southern Olympic Mountains over the next 50 years. Western hemlock, western red cedar, and whitebark pine, on the other hand, may expand their ranges across the Pacific Northwest by the end of the century.

For all trees at higher elevations, warmer temperatures that reduce snowpack will induce both an earlier snowmelt and growing season, conditions that will likely lead to an increase in tree establishment and growth. Warmer and drier conditions may introduce new risks from insects, disease, and invasive species, but this is still speculative. Warmer temperatures and less moisture may increase fire activity in the Puget Sound region. Without an extensive fire history to learn from, it is unclear how fires may affect the area's forests. One thing that remains clear is that continued warming will bring changes to the makeup of epiphyte and old growth forest communities.

RIDDLES OF THE FOGBIRD

I am wide-awake at one a.m. The alarm won't sound off for another hour, but I am eager for today's work. I plan to head up the valley, get into position an hour before sunrise, and wrestle with one of North America's most baffling ornithological puzzles—the nesting place of the marbled murrelet. The mystery so stumped scientists and birdwatchers for 185 years that in 1970, the National Audubon Society offered a cash prize for "the first verified and documented discovery of the nest of the marbled murrelet." Ornithologists knew where the other 700-plus North American birds bred, but the nest of the short-necked, stubby-tailed, chunky, robin-sized seabird eluded searchers for nearly two centuries. Murrelets frequent Washington's nearshore marine waters outside the breeding season, yet we have known for less than 50 years that they raise their young in old growth forests.

Murrelets reach their peak of inland activity in July, using roads and rivers as flight pathways. Younger birds and non-breeding adults accompany breeding adults on the daily, 90-mile round-trip from the Salish Sea to the Carbon River forests. The young and non-breeders may be prospecting for future nest sites; breeding adults trade incubation duties with their mates or bring food to nestlings. Most activity occurs around sunrise, the best time to see or hear the birds. They return to saltwater each day, with only the incubating adults staying behind for their 24-hour shift on the egg.

Besides the inconvenient time frame for observing marbled murrelets, other things complicate matters. The dense forested habitat, the bird's small size and sooty brown breeding plumage, its tendency to nest solitarily, and its activity during periods of low light make the birds extremely difficult to see. Their proportionately long and narrow pointed wings and rapid wingbeats allow them to fly at high speeds averaging between 45 and 85 miles per hour. Some have been clocked at nearly 100 miles per hour. These traits make the birds more easily heard than seen. Only 20 percent of the breeding season observations are visual—most come from hearing its call. In spite of the graveyard shift and the lousy odds of actually spotting the bird, I resolutely head up valley with a friend and fellow murrelet chaser. We are up before even the robins and varied thrushes have begun their dawn chorus. Our headlamps light the way. We turn them off at the first hint of daylight to savor the challenge of groping our way along the trail. We have studied the bird's call and know what to listen for. Like most foolhardy souls, we are hopeful.

Once on the river with thermoses in hand, we settle in and wait. The river mumbles groggily while robins, thrushes, and tanagers tune-up. A pygmy owl toots rhythmically off in the distance. Raven squawks the sun up into the sky. A pesky cloud of no-see-ums threatens to evict us. We are eventually rewarded with one audio observation, a *keer* sounded five times in rapid succession. Though distant, we easily distinguish the murrelet's gull-like call.

Scouting the river's edge for possible nest sites, I found a stand of four hulking Douglas firs, the tree most commonly used for nesting. They towered over the neighboring hemlocks, cedars, and spruces, each with a diameter of six feet or more. With binoculars, I carefully scanned the near-trunk, middle to top-third of each tree, the usual nest location. Hefty limbs with mossy platforms looked promising, but the tangle of branches obscured my view. Suddenly, a shadowy, feathered missile zoomed beneath the canopy directly in front of my companion's outpost, an indication of possible breeding activities. We were elated with our good fortune: two novice observers had heard and gotten a rare glimpse of the marbled murrelet. On seven subsequent trips, we recorded from 13 to 48 observations per visit, each ranging from one to eleven *keers*.

The marbled murrelet is known to ornithologists as *Brachyramphus marmoratus*. *Brachy* means "short" and *ramphus*, "curved beak." *Marmor* means "marble," for the brown and white marbling coloration of breeding adults. Native Alaskans and old-time loggers called them fog larks or fogbirds, presumably because of their calls and the way they disappeared into the early morning coastal mist. They belong to the family Alcidae—seabirds that include the puffins. Marbled murrelets spend most of their lives at sea but unlike their relatives that nest colonially on cliffs or in burrows, they are the only seabird that nests in old growth forests. Like the anadromous fish that spend part of their lives at sea and return inland to breed, they link marine and terrestrial environments, which presents them with many challenges.

Common in the Aleutian archipelago, the birds breed from Alaska to central California. In Washington they occur on Puget Sound, the Strait of Juan de Fuca, and on the outer coast. Near Neah Bay at the north-westernmost point of the Olympic Peninsula, archaeologists at the Hoko River rockshelter found skeletal remains of murrelets from 2,500 years ago. They also found evidence a short distance south at Ozette, another Makah Indian ancestral site. Further north, in the Yakutat area on the Gulf of

A marbled murrelet incubates its lone egg, set in a mossy hollow on a large limb, high in a Douglas fir. *Kirsten Wahlquist*

Alaska between Anchorage and Juneau, the birds held spiritual significance for the Tlingit people. Stories tell that "raven put his mother in its [the murrelet's] skin during the flood" and that the Tlingit do not eat the murrelet because "it was raven's mother."

At Mount Rainier, trained observers conducting audio-visual surveys and others using marine radar have located marbled murrelets in the Nisqually, Mowich, South Puyallup, and Carbon River drainages. Radar surveys (on which the birds present a unique flight signature) indicate that this valley supports the highest density of birds within the park. To date, no one has found a nest within park boundaries, but the bird's annual presence indicates probable nesting.

Early clues to possible marbled murrelet nesting sites came in the late nineteenth century when observers saw them fly inland from the Alaska Coast near Sitka. Alaskan Natives, probably Tlingit or Haida, said that the bird nested in hollow trees up in the mountains. Within a decade, other observers spotted birds flying inland from the Washington Coast. Native Americans near Mount Baker told of them nesting high in the Olympic Mountains. People found eggs, chicks, and dead adults in old growth forests in the first half of the twentieth century, likely from falling out of trees cut by loggers. Searchers discovered ground nests in Alaska in the 1930s. Finally, in a California state park in 1974, a tree surgeon came face-to-face with a nestling nearly 150 feet up a massive Douglas fir. The beleaguered bird tumbled out of the nest and fluttered to the ground. It died the next day, but scientists preserved the specimen and nest as proof of the first well-described tree nest of the marbled murrelet.

With over 250 old growth nests now recorded (including 20 located by a radio tagging study in Washington), researchers have developed a clear picture of the bird's nesting habits. Instead of building nests, they use shallow depressions in moss-covered platforms on large-diameter limbs. They need vertical cover above the nest cup for protection from the elements and from predators, and horizontal access for entering and exiting. The female lays a single egg as early as March, when the 28- to 30-day incubation period begins. Hormonal changes in both the female and male create a brood patch, a featherless area on the birds' bellies. A network of blood vessels near the skin's surface helps regulate the egg's temperature. The parents take turns incubating the egg, with one foraging at sea while the other broods. The bird remains motionless upon the egg for more than 90 percent of the time. Partners exchange roles at dawn each day.

The chick is born down-covered and alert, unlike the naked, blind, and helpless condition of most songbirds. It sleeps most of the time, and is vulnerable to predation and falls from the tree, especially during its first week. The parents feed their rapidly growing offspring from one to eight times per day, each carrying one fish at a time crossways in its bill. They feed on Pacific sand lance, Pacific herring, or northern anchovy, seeking the larger but harder-to-find fish in order to minimize the number of feeding trips. Nests have been located nearly 37 miles inland in Washington, meaning that some birds fly a minimum of 73 miles per day to feed their young. Suitable nesting habitat lies as far as 55 miles inland, and considering that most birds make multiple daily trips, it's likely that some fly hundreds of miles each day.

As the 28-day nestling period winds down, the young bird increases its wing flapping activity and removes the last of its down by scratching and preening. Now sporting its black and white juvenile plumage, it's time to fledge. Alone, at dusk, and without so much as a single training flight, the fledgling leaves the security of the nest and begins its perilous nonstop journey to the sea. Falling short of the target and becoming grounded is deadly; regaining flight altitude from land is nearly impossible. Like its ancestors, the bird uses creeks, rivers, and sometimes roads as giant compass needles to guide it to the ocean. If it has the strength and endurance to find its way, while successfully evading predation on its inaugural flight to coastal waters, the newly fledged bird alights within about a mile of shore. Without any further contact with its parents, the bird soon begins foraging on its own. It uses its wings to "fly" underwater as it chases prey to depths of over 50 yards, often remaining submerged for more than half a minute. Come winter, the young bird's diet will switch from small schooling fish to krill and other shrimp-like crustaceans. It learns to elude bald eagles, peregrine falcons, western gulls, and northern fur seals. It adapts to life on saltwater, forging one of the most critical links in its species' chain of survival.

The marbled murrelet spends 95 percent of its life at sea with the remaining time spent on land during the breeding season. This dependence on disparate ecosystems exposes the birds to a variety of environmental hazards that threaten their survival. As nearshore marine feeders, they are especially vulnerable to oil washing ashore from spills. The 1989 *Exxon Valdez* disaster, for example, killed at least 350,000 seabirds. Of those, 8,400 were murrelets, most of which were marbleds. Smaller spills and incidental dumping—often legal—kills scores of birds every year. Another hazard at sea is the Alaska gill net fishery that annually nets millions of salmon. The nets trap and kill up to 3,300 birds each year.

Although marbled murrelets spend only a fraction of their lives on land, the impact of land-based human activities has devastated their population. The timber industry began logging old growth forests of Washington, Oregon, and California in the late nineteenth century and continued well into the twentieth, shrinking them from 66 to 10 million acres in the span of two human lifetimes. This left the birds with just 10 percent of their historic nesting habitat, threatening their ability to maintain stable populations. As a result, the murrelet gained protection as threatened under the Endangered Species Act in 1992, two years after the spotted owl listing. Oregon and Washington state-listed it as threatened in 1993, the same year that California classified it as endangered.

Recognizing the link between murrelet numbers and nesting habitat, the U.S. Fish and Wildlife Service designated 3.69 million acres of old growth forest as critical habitat, giving the bird special protections. Much of the remaining habitat, however, lies in relatively small and fragmented tracts. One consequence of these unconnected forest patches is large edge-to-area ratios that make nests more visible and accessible to predators. Despite the murrelet's elaborate physical and behavioral traits intended to thwart predators—cryptic coloration, rapid flight, and well-hidden nests with minimal activity restricted to periods of low light—78 percent of their nest failures are the result of predation.

Pacific Northwest marbled murrelets are the target of more than a dozen bird and mammal predators, but corvids stand out as the most common and pernicious. Common ravens, American crows, Steller's jays, and gray jays, all of them audacious, interloping omnivores, seize eggs, young, and adults whenever the opportunity arises. Campgrounds and similar areas attract food-scavenging corvids, which can be fatal for birds nesting nearby. Rangers at some California recreation areas have used the slogan "Feed a jay, kill a murrelet" to educate and induce visitors to refrain from feeding wildlife.

Considering a clutch size of one, a life cycle brimming with danger in two ecosystems, and the catastrophic loss of nesting habitat, it is no surprise that marbled murrelets have precariously low reproductive success. Their low rates of hatching success, fledgling survival, and recruitment of young birds into the population imperil the bird's future. A study in Washington found that only 13 percent of tagged birds attempted to nest and that just 20 percent of those attempts succeeded, leading biologists to doubt whether adult birds can replace themselves in the population.

In 2016, wildlife biologists Gary A. Falxa and Martin G. Raphael released a Northwest Forest Plan report on the bird's nesting status and its population trends. Data showed a 27 percent loss of nesting habitat on state and private lands over a 20-year period, nearly all of it due to timber harvest. The 276,000 acres of lost habitat equals an area three times the size of Seattle. In terms of population trends, boat-based surveys on the coastal waters between Washington and northern California collected data between 2000 and 2013. The good news is that murrelet numbers appear stable in Oregon and California, but the data for Washington birds is alarming. Evidence of a 4.6 percent population decline per year concerns Falxa, Raphael, and their colleagues, who believe that the downturn warrants action. The report found a correlation between the area where numbers had dropped the most—on Washington waters—and the greatest overall loss of nesting habitat. To help remedy the trend, the researchers suggest that landowners need incentives to willingly reduce fragmentation, create edge buffers, and preserve high-quality blocks of forest interior nesting habitat.

Census results from the Washington Department of Fish and Wildlife aligned with the regional report, with at-sea monitoring showing a 44 percent drop overall in the Washington marbled murrelet population between 2001 and 2015. With a current estimate of about 7,500 birds on state waters, the evidence was so startling that the department recommended changing the state's listing of the bird from threatened to endangered.

Outside the Washington-Oregon-California region, the birds are faring no better. Numbers have fallen off in Alaska by about 70 percent in 25 years, where they are most numerous. Canada lists it as threatened and reports a similar pattern. In addition to continued census surveys, researchers need to study second growth forests to understand whether the birds can nest there. Other projects can examine the relationship between the murrelet and its prey species. As climate change elevates sea surface temperatures, the impacts on forage fish may affect the birds in multiple ways.

The mysterious marbled murrelet is an anomaly among seabirds. "Mother of Raven," its nest site remained hidden for generations among the mightiest conifers, many miles from its food source. Without parental aid, the fledgling makes its first trip to the Salish Sea. An uncertain future lies ahead, the result of dwindling habitat and declining numbers. Only fast action will stabilize, and eventually increase, its numbers range-wide.

AND THE PEOPLE CAME

Along with hiding the nests of the marbled murrelet, the lower valley's dearth of direct sunlight, dense forest shrouded in mist or drizzle, and an impervious undergrowth combine to obscure most traces of human presence through the ages. Although both Muckleshoot and Puyallup tribal people hold long-standing ties to the corridor that lies within their traditional use areas, thickets of devil's club and the many downed trees make it extremely difficult to search off-trail for archaeological evidence. Today retired park archaeologist Greg Burtchard, an expert on Mount Rainier's cultural resources, and I will explore some of the Carbon's upper elevation sites that date back thousands of years.

We depart Mowich Lake for Spray Park to a Steller's jay's raucous call. Further on, a hairy woodpecker's mad staccato hammering telegraphs its message: "This way to the high country." After a couple of hours, we make the lower section of Spray Park. This elegant, late August morning boasts cottony strands of clouds that frame the mountain in a gloriously surreal way. The craggy, foreboding peaks of Echo and Observation Rocks serve as a backdrop for the sprawling meadows. The height of the wildflower season passed a few weeks ago, but the splendid, deep blue patches of mountain bog gentian add striking contrast to the green-hued meadows. Gentian blooms late in the season, signaling autumn and a coming change in the weather pattern.

One of the first archaeological finds at Mount Rainer, a projectile tip, was found in Spray Park in 1930. It was made of chalcedony, a type of quartz that includes onyx, agate, and jasper. The first people to venture into Tahoma's subalpine parklands worked chalcedony into projectile points and other stone tools. The craftsman had flaked the stone on both sides, giving it what archaeologists call a biface tip.

Burtchard points out a stone pit feature discovered during field reconnaissance in the early 2000s. The pit lies subtly in the landscape, invisible to most passers-by. It is the only one found to date at Mount Rainier and experts are uncertain regarding its purpose, but such pits are common east of the Cascades crest.

On the northern edge of Spray Park and facing east above Mist Park, we spot a rockshelter at the base of Mount Pleasant. Looking like a dark eyebrow etched into the cliff's haggard face, it contained rock flakes from indigenous toolmaking or maintenance. Wood fragments and blasting caps linked the shelter to historic times, when it was probably used by prospectors.

We stop for lunch at a spot that overlooks the vast green subalpine meadows of Mist Park, a mosaic of seasonal puddles, pools, and rivulets. Burtchard explains that grouse gizzard stones found at this site indicate that people probably roasted and ate the birds here hundreds to several thousand years ago. Stone flakes from sharpening and maintaining tools signal use of the spot as a short-term camp, frequented only during the snow-free summer months. The camp's sweeping vantage point allowed hunters to scan for elk in the meadows below and for mountain goats on the hillsides above.

Other sites help prove the presence of first people in the upper Carbon River watershed. Although lacking high quality rock for toolmaking, several small quarries provided useable material. Local sites included one at Mist Park, one near Mowich Lake with a greenish chert (a sedimentary rock resembling flint), and a third site around Knapsack Pass with lavender-colored chert. Projectile points found on the trail near the Tolmie Peak Lookout and below it at the west end of Eunice Lake give more clues about first peoples' hunting trips. Geologist Bailey Willis—who in the late 1880s spent considerable time searching for coal in the area and later became a staunch advocate for the creation of Mount Rainier National Park—reported finding Indian baskets near Eunice Lake. Taken together, these sites and their associated finds place indigenous people in the watershed for at least 4,000 years. With more fieldwork, it is likely that researchers could push the clock back even further.

The Native Americans who frequented this valley sought the same things that people on other sides of the mountain valued: mountain goat, marmot and other animals, huckleberries, bear grass, and other plants. The first European Americans to venture up the Carbon pursued different resources. Following the discovery of coal seams in 1875, the mining towns of Fairfax and Carbonado sprang to life. Companies began extracting metal ore from within park boundaries by the turn of the century. Two mineral ore mills—mostly unsuccessful—operated in the valley until 1909 and 1913, respectively. Spunky tourists had begun visiting the area in the 1880s, and with the end of mining activities, park staff hoped to make the Carbon River a destination for visitors.

Early in the twentieth century, a primary challenge for visitors was getting from Seattle and Tacoma to the park and once there, traveling through it. Wagons and railroads brought hundreds of people annually, most of them to Longmire Springs. The other popular attractions were Paradise and Indian Henry's Hunting Grounds. The Nisqually-Paradise Road connected the southwest entrance with the Paradise meadows, but many park advocates also sought an easier way to access the Carbon Valley. Instead of traveling by train and then on foot or horseback, they wanted a road into the park's northwest corner. The area lay closer to Seattle than Longmire or Paradise, and many felt that the Carbon Glacier's terminus rivaled Rainier's other scenery. The larger vision was to tie the motorway to the "Around-the-Mountain-Road," an ambitious, grand plan of the early twentieth century to build a road around the entire mountain just below the glacier line. Although later abandoned in favor of a more practical reliance upon roads outside the park, a survey for the Carbon River road began in 1915.

From the outset, designers saw it as a pleasure drive along the river blending in seamlessly with its natural surroundings. Instead of the sweeping panoramic views of the road to Paradise, it would nestle among behemoth old growth trees and lush vegetation as it followed the riverbed to the glacier. Funding shortfalls delayed the start of construction until 1921, when trouble soon befell the project. Flooding caused damage while the project was still in its early phases, a troubling omen of things to come. Despite the setbacks, the eight-mile road from the northwest entrance to the glacier's terminus opened in 1924. The Ipsut Creek Campground, five miles up the road, opened in 1925.

The finished road delivered the motoring experience that the engineers had envisioned, but they had overlooked some crucial environmental factors. The route along the path of least resistance and wagon road technology that failed to use modern engineering techniques created serious issues. Like most of Mount Rainier's rivers, the Carbon is a braided, high-energy, disturbance-prone system characterized by heavy sediment loads. Unstable channels that contain sand, gravel, cobbles, and boulders shift frequently. Water often overtops the banks. Federal experts examined the road after its completion and concluded that its alignment in the floodplain and the failure to recognize the river's tendency to change course would consign it to an existence of flooding, washouts, and repairs.

Their assessment proved prophetic the next year when water damaged the road and necessitated the construction of embankments intended to prevent future problems. The Ipsut Creek Camp of the Civilian Conservation Corps (CCC), one of seven camps of up to 200 men each at Mount Rainier between 1933 and 1940, poured thousands of hours of labor into protecting the road. They maintained, repaired, and improved it by installing log cribbing and rock gabions to stabilize the riverbank, redirect the river, and prevent flooding events. Despite the CCC's optimistic slogan—"We can take it!"—the corpsmen's efforts were no match for the powerful river, and rockslides and flood damage continued through the 1930s and 1940s. By 1950, it became clear that the river would continually have its way. Park staff closed the last three miles of road between Ipsut Creek and Cataract Creek and converted it to a trail.

Floods and washouts continued over the next 40 years. The most devastating was a 100-year flood in the spring of 1996, when a portion of the rampaging river flowed directly down the road and gouged out a deep trench. It destroyed a dike and a picnic area, causing damage estimated at $500,000. When the road reopened two years later, the next flood completely obliterated the restored section.

The travails of the Carbon River Road came to a head in November 2006 with a record-breaking downpour that dumped 18 inches of rain in 36 hours. The river overran sections of road, damaging six spots that totaled over one linear mile. It was the latest of at least a dozen different incidents that rang up repair costs between $1.3 and $1.6 million, and it was time for desperate measures.

Despite its troubled past, the road was not abandoned. First, it provided access and generations of memories to visitors who loathed the idea of losing easy passage into the Carbon basin. Second, it receives federal pro-

The Civilian Conservation Corps built rock and log cribbing structures along the Carbon River in the mid-1930s, trying to prevent flood damage to the road. *Courtesy of Mount Rainier National Park Archives*

tection as a cultural landscape within the Mount Rainier National Historic Landmark District. Along with other roads and historic buildings, land managers are obligated to preserve and maintain historic features. The road is noteworthy as an early example of a scenic highway and it remains one of the few unpaved drives of its kind left in the National Park Service system.

With input from the public, park officials embarked on a project to convert the historic road to a trail that preserved hiking and bicycling access without the cost of rebuilding and repairing the road for automobile traffic. The proven success of engineered log jams (ELJs) for erosion control and habitat enhancement, as demonstrated in the Nisqually River drainage, made ELJs the best choice for the project. In tandem with 10 log check dams that slow river flow, the five logjams stabilize the remaining riverbank, create a place for sediments to accumulate on the riverbed, and divert the river away from the road.

The other goal of the $3.2 million project included habitat enhancement for the federally protected bull trout, and that's where things get tricky. Logjam installation must proceed without any impacts on these fish, which are genetically distinct from those in Mount Rainier's other drainages. Ironically, the best way to respond to the disturbances created by the river is to create another disturbance. Installing check dams and logjams requires the use of heavy equipment that moves tons of logs and other debris, temporarily redirects river flow, and threatens the lives of thousands of riverine organisms. To mitigate the potential effects and to follow strict U.S. Fish and Wildlife regulations protecting the trout, the park's aquatic technicians work nonstop alongside the excavator, recording the river's temperature, pH, dissolved oxygen, and other water quality parameters. Most important is turbidity, the amount of sediment suspended in the water. If the equipment stirs up too much sand and silt and the turbidity exceeds allowable levels, the work must stop. Another concern is the noise that the thundering machine generates and its potential effects on the marbled murrelets that may be nesting nearby. A technician monitors the ambient noise levels to ensure that they stay within federal standards.

The hydraulic excavator's first task is to reduce water flow near the road in order to install the logjams. It does this by moving bucket after bucket of fill to create a diversion channel to carry the water running parallel to the road into the river's mainstem. Within a few hours, the machine has moved enough gravel, cobbles, and logs to de-water (empty) the side channel and allow logjam installation to begin. Most of the river's inhabitants—the seldom seen fish, amphibians, and invertebrates—move away quickly as the bone-jarring machine lumbers about. Salamanders and other aquatic animals burrow into the river bottom to wait out the disturbance, but many are less fortunate. Puddles left behind teem with life. If they dry up, the organisms will die. Technicians, student interns, and volunteers quickly spring into action, hand scooping fish and amphibians into five-gallon buckets to count, measure, and then release into the fast-flowing mainstem. "We're saving lives!" shouts a biologist. Our buckets soon fill with sculpins, a minnow-sized bottom feeding fish, the number one food of bull trout. Even with the precautions, mortality mounts in the now-puddled channel. Some fish will inevitably die in the quickly shrinking pools. Masses of stoneflies, mayflies, and caddisfly larvae sit high and dry on river rock.

Opportunistic American crows and a spotted sandpiper crowd around one wet spot, picking off stranded sculpins and exposed invertebrates. An American dipper—usually seen in fast-moving streams and rivers—soon joins them. These chunky, sooty gray birds that stand a bit smaller than a robin are unlike any other North American songbird. Common, though easy to miss, they are our only aquatic passerine (perching birds that include all of the songbirds). Dippers walk, dive, and use their stubby wings as flippers as they forage underwater for aquatic invertebrates and the occasional small fish or fish eggs. Their strong legs and toes, combined with their counterbalancing wings, enable them to withstand river currents strong enough to sweep a person off their feet. Thick down, an abundance of contour feathers, and a low metabolic rate equip them for a life in cold mountain waters.

One of the bucket patrollers finds a northwestern salamander, so we gather round to briefly admire and study it. Brownish and about seven inches long, this adult is a paedomorph, retaining large, feathery, external gills from its larval stage. The northwestern salamander is one of three salamander species in the park in which paedomorphs occur. Some adults metamorphose into a non-gilled terrestrial form and move into the forest to live. Northwestern salamanders are the top aquatic vertebrate predator in many subalpine lakes and ponds (and in lowland fishless lakes and ponds) within their range and are widespread and abundant at Mount Rainier.

We soon find a coastal tailed frog in another pool. Endemic to the Pacific Northwest, it occurs nowhere else in the world. Considered the most primitive living frog, its tail is actually a cloacal protuberance that helps internally fertilize females. It is the only genus of North American frog adapted to life in frigid waters.

The northwestern salamander and coastal tailed frog are two of the 14 species of amphibians found within Mount Rainier National Park. Amphibians are common but most species are difficult to observe because of their nocturnal and secretive habits. The Cascade frog and the rough-skinned newt are probably the easiest to see in the park. Amphibians' evolution from the fishes more than 360 million years ago and their subsequent movement from water to land marked a breakthrough in vertebrate evolution. Derived from Latin and Greek words meaning "living in both water and on land," amphibians truly lead a double life. Their two-phase life cycle generally includes an aquatic larval stage while many adults live on land.

Amphibian presence indicates a stream's relative health, and an increase in water temperature, siltation, or decreased oxygen levels can endanger them. Because of the potential threats and the great number of individuals and species present at Mount Rainier, aquatic researchers closely monitor their populations. A long-term alpine lakes study tracks water quality at six mountain lakes within the park. On amphibian visual shoreline surveys, technicians carefully walk the lakeshore to make and record observations. During one survey, I watched 10 northwestern salamanders dart for cover after a researcher dislodged the rock under which they were hiding. Once he identified them and recorded their numbers, he meticulously returned the rock to its exact spot. At another lake, I recorded data while Scott Anderson, the park's amphibian specialist, conducted a nearshore snorkeling survey. Wearing a rubberized dry suit that suggests a kid-friendly, scientific version of the monster from the 1950s horror film "Creature from the Black Lagoon," Anderson swims along the lakeshore in search of amphibians. A ruler for measuring specimens and a waterproof camera and flashlight complete his outfit.

Nighttime surveys during rainy periods allow Anderson and his co-workers to map amphibian presence on park roads. The data can be useful in planning maintenance and construction work that might otherwise impact areas with high amphibian densities. Amphibians outnumber cars on most night surveys. Our best find was a nine-inch long coastal giant salamander, its girth as great as a child's wrist, poised stock-still on the roadway.

When we finish transferring stranded fish and amphibians from the shallows to the mainstem at the Carbon River ELJ project, some of the aquatics crew use seine nets to catch fish hiding in the deeper puddles and pools. In chest-high waders and working in pairs, they quickly net the next round of holdouts and get them into the river. When finished with the nets, they begin the last phase of fish finding. Called electrofishing, this common method stuns the fish in order to net and then move them. One technician wears the battery unit while holding a pole in the water. When the 14-inch aluminum hoop on the pole's end receives current, it stuns all fish in the vicinity. Other staff deftly net and quickly relocate them. Although electrofishing typically results in little or no permanent harm to the fish, occasionally there's some mortality. For every fish that dies, however, many are rescued, a compromise that the biologists are willing to make.

Once the crew finishes their salvage efforts, the excavator can begin building logjams at sites ravaged by floodwaters. Where the bank is badly undercut, many trees lie toppled into the river while others hang on precariously. The equipment operator buries most of a 30-foot log vertically into the riverbed, leaving 10 to 12 feet of it exposed above ground. After burying several logs this way, he then wedges root wads, smaller logs, and other debris between the uprights, building a structure that will slow the water's flow, allow it to drop some of its sediments, and protect the riverbank from further erosion. After finishing the work, the excavator restores the water flow to its original path and backs out of the floodplain, covering its tracks the best it can. There's nothing else to do now but wait for the weather to turn, the rains to come, and see if the logjams do their job.

Returning to the site nearly a decade later, I find hundreds of "volunteer" red alder saplings blanketing the tops of the logjams and the neighboring area. A few Douglas firs, Sitka spruce, cottonwood, and willow fill in. Red-legged frogs have bred at the base of one logjam. The best news is that the ELJs remain in place, veterans of multiple floods. They've held their ground and even better, re-routed the river away from the road. Though other sections remain at risk, this one is secure. It's counterintuitive, but for now, the idea of "further disturbing the disturbed area" appears to be working.

A Stack of Plates

In the pre-dawn hours, I drive around to the mountain's southwest side for a hike up Pyramid Peak. Looking up from the road and toward the mountain, I gape at a complex cloud formation just west of the summit. It is an unusual grouping of six layered, pancake-like clouds in an inverted triangle in shades of cream, ivory, and gunmetal.

Known as lee wave clouds or *pile d'assiettes*, French for "stack of plates," these ethereal beauties are the occasional products of a mountain wave cloud, a hallmark of the region's montane skies. Mountains play an essential

Lee wave clouds, a type of lenticular cloud, hover over the mountain. *Photo by Kevin L. Bacher*

role in their formation and when it's an imposing and isolated rampart such as Mount Rainier, the smooth-looking results are often lens-shaped. Called *altocumulus lenticularis* by meteorologists and known as lenticular clouds, they generally reside in the mid-atmosphere between 5,000 and 20,000 feet.

Lenticular clouds are credited with launching the unidentified flying object (UFO) craze that received widespread press coverage beginning in 1947. In June of that year, Boise entrepreneur and pilot Kenneth Arnold reported seeing nine saucer-like objects near Mount Rainier while flying from Chehalis to Yakima. Subsequent media reports turned his sighting into "flying saucers," giving rise to UFO fever. Observers reported over 9,700 sightings in the United States between 1947 and 1965.

Mountain wave clouds form when an air parcel flows up and over mountainous terrain. After its passage over the peak, the air continues oscillating vertically, much like a car with worn shock absorbers that continues bouncing up and down after hitting a bump. If the atmosphere contains alternating layers of moist and dry air, clouds can form in the moist layers at the wave's crest. There the cooler air condenses the water droplets, forming the "plates." Clouds don't form in the wave's trough where the air is warmer. These are the layers between the plates. More common are cap clouds. These form directly over the summit when winds push water vapor up to the crest. There it cools, becomes saturated with moisture, and forms a cloud. Air flows freely through them; they can stay put for several hours. Seasoned outdoor travelers know that lenticular clouds, whether lee wave or cap, signal an approaching weather system. They indicate winds whipping up to 20 to 30 miles per hour and increasing humidity, conditions that usually prefigure precipitation. Rain or snow usually follows within 24 hours, but occasionally the front veers to the north or south.

A series of scarf-like cap clouds dominated the mountain during my hike and appeared intermittently for the next few days without producing a broad band of wet weather. When they finally delivered on their promise of moisture, I knew exactly where to go: back to the Carbon. My last five or six trips there had been rain-free, and it was time to appreciate the area in its dank, dripping glory.

Heading up the historic road, raindrops land like cold steel pinpricks on my bare neck. Puddles dot the road, some circular, others oblong or teardrop. I peer into them. A few are opaque windows, cloudy with sediment. Others are mirrors, reflecting my silhouette back at me.

Eventually the rain stops and the sun appears. Shafts of light angle through the trees onto the road. Tiny billows of steam swirl and rise from damp vegetation. Water droplets in the hanging mosses gleam and twinkle like radiant jewels. All is clean and renewed.

Back at the entrance gate a bit later, the sun slips away. The rain resumes. The clouds surge up valley, an escalator of moisture. They release their precious cargo before the drier air descends to Yakima Park and Sunrise on the northeastern slopes. Without raingear, I moisten up one more time. This is exactly what I had hoped for, to revel in the Carbon River valley's true spirit: Rain. Respite. Rain.

7

The Sunrise Area:
The High and Dry East Side

Mount Rainier from the Sunrise area; Little Tahoma Peak is far left. *Drawing by*
Lucia Harrison

OF NUTS AND NUTCRACKERS

I arrive at sunrise on the park's northeast side and immediately begin
searching for Clark's nutcracker, my favorite subalpine bird. Jay-sized
and appearing to sport a classic tuxedo, there's no sharper-dressed bird on
the mountain. Its medium-gray body, black wings with white patches, and
black tail with white outer-tail feathers sets it apart from the less flashy
gray jay. They inhabit these higher areas, but winter at lower elevations.
These early nesters build a well-hidden, protected nest in a conifer as early

as January, more typically in March. Like marbled murrelets, both parents develop a brood patch. This allows each of them to incubate the eggs while their mate forages for food. After the young fledge in the spring, the entire family moves up the mountainsides for the summer and fall. Early breeding enables their three to five young to become independent by late summer when they begin caching food for winter. Visitors often call nutcrackers and gray jays "camp robbers" because they can become food-conditioned scavengers, taking handouts from careless picnickers and campers. Their natural diet includes insects, carrion, and pine seeds.

Clark's nutcrackers are unusual in that they have a mutualist relationship with whitebark pine that benefits both species. The tree ranges at elevations between 3,800 and 7,000 feet in a broad band extending from Canada south and east through the Olympic and North Cascade Mountains into California, Idaho, Montana, and Wyoming. Most occur in remote locations that include wilderness and roadless areas, with nearly all of them on public lands. To hike, climb, camp, or ski among whitebark pine is to be in the high and wild backcountry.

Whitebark pines develop cones and seeds only after 100 years and can live for 700 years or more. Nutcrackers depend on whitebarks as a primary nutritional source, obtaining vital fats, carbohydrates, and protein from its seeds. The tree relies on the bird to collect and disperse its seeds in order to regenerate and extend its range. The wind can't transport the large, wingless seeds, but the bird distributes them over a greater distance than breezes carry the seeds of its top competitors, Engelmann spruce and subalpine fir. If you spot a Clark's nutcracker, you can be fairly certain that whitebark pines grow nearby.

The bird begins collecting seeds in midsummer, using its sturdy, sharp bill to break the scales from the closed cones to expose the seeds for extracting. It removes the seed, then shells it by cracking it in its bill or hammering it while holding it in its feet. It then places the seed in a pouch under its tongue, the only such storage structure known to any bird. The pouch has a capacity of about one and a half tablespoons and holds an average of 70 seeds. With a full pouch, the nutcracker flies up to a few miles and as much as 2,000 feet higher on the mountain to cache the seeds to eat later. Each bird stores as many as 98,000 seeds per season in stockpiles that hold three to five seeds about an inch deep in the ground.

The bird's remarkable spatial memory allows it to use landmarks like prominent rocks or trees to find seed caches up to nine months after bury-

A Clark's nutcracker in a whitebark pine near Dege Peak. *Walter Siegmund, Wikimedia Commons*

Hoary marmots lounge alongside the Skyline Trail in the Paradise area. *Joe Mabel*

ing them. In years of large seed crops, the birds harvest and store far more than they need for food. Unrecovered seeds may germinate to become seedlings. Barring environmental disturbances, they establish themselves and eventually the cycle is repeated.

I once watched a pair of Clark's nutcrackers working among whitebark pines on the trail between Frozen Lake and Berkeley Park. One flew into a tree's crown, examined a cone, worked it vigorously with its beak, dug out a seed, and moved on to another tree. The second bird repeated the sequence. Silently I crept after them, binoculars up, amazed by their stealth and efficiency.

Other animals share in the late summer bounty of whitebark pine seeds. Jays, ravens, chickadees, and nuthatches also eat them. Ground squirrels and chipmunks harvest and store cones in middens that attract black bears in the Cascades and grizzlies in the Rocky Mountains. At Yellowstone, grizzlies feed almost entirely on whitebark pine seeds in the fall. When the seeds are plentiful, human–grizzly interactions are fewer because the bears are busily foraging at higher elevations further away from developed areas and park visitors.

But whitebarks provide more than just food for wildlife. Animals also use the trees for shelter, nest, and burrow sites. The pine's ability to withstand temperature and wind extremes enable it to colonize exposed, inhospitable ridges where it often acquires a gnarled and windswept posture known as *Krummholz*. Once established, it influences snowmelt patterns and water runoff. Communities of fungi, lichens, mosses, and other plants thrive under its protective cover to add structural complexity and biodiversity to the understory. This influence beyond its size or numbers makes it a keystone species, which in the absence of disease or pests, has far-reaching effects. Several factors threaten its long-term survival, however, which could create potentially devastating effects on subalpine systems. That story begins with human interference, starting over a hundred years ago.

TROUBLE IN THE HIGH COUNTRY

On an otherwise ordinary day on the docks at Vancouver, British Columbia, a shipment of eastern white pine seedlings arrived from a European nursery. Originally exported to Europe because of its value to the timber industry, growers there often sent the trees back for use as ornamentals. A stowaway on the seedlings, a fungus native to Asia, was about to wreak havoc on five-needled pines in western North America.

Known as white pine blister rust, the fungus arrived on the West Coast in 1910 and spread quickly throughout the northwestern United States. It reached the southwestern section of Mount Rainier National Park by 1928, where it appeared on currant shrubs and western white pines. A blister rust control program that began in 1930 removed hundreds of fungus-hosting gooseberry and currant plants but did little to stop the pathogen that reached the Sunrise area by 1937. Within 14 years, over half of Mount Rainier's whitebark pines showed signs of infection. The quick, destructive pace of the fungus continues in the Pacific Northwest, where it's now considered widespread and common. Biologists expect wholesale mortality in the coming decades.

The fungus takes its name from the thimble-sized pouches on the tree's bark that resemble orange-yellow blisters. Tiny airborne spores are able to drift by the millions over 300 miles to land on the leaves of its hosts. The new arrivals transform into other spores that spread to nearby shrubs. During cool summer nights, a third spore type forms. When the relative humidity holds close to 100 percent for six to eight hours, they germinate and produce yet another kind of spore, which land and germinate on moist whitebark pine needles, infecting the tree. Thread-like filaments called mycelia enter the needles through the stomata, the needle's tiny underside openings that permit gas exchange. The fungus moves through the needles into the branches at a rate of about four inches per year to infect the tree's upper cone-bearing branches. Seed production ceases. When the fungus encircles the trunk, it kills the tree. Whitebark pine mortality at Mount Rainier is at about 31 percent. The infection rate seems to increase from west to east and from lower to higher elevations, putting the trees at Sunrise squarely in the path of the devastating fungus.

The good news is that blister rust rarely wipes out all the pines in an infected stand, indicating some genetic resistance to it. Trees without cankers or those cankered but still alive have rust-resistant genes that may signal hope for the species' survival. Collection and propagation of disease-resistant seeds hold promise for its future. Still more good news is that of the thousand Mount Rainier saplings tested, 62 percent were uninfected. Lab tests found the park's seed to be more blister rust-resistant than specimens from anywhere else in the region. Experts are concerned though, that if blister rust runs its course, the tree will undergo local extinctions with potentially disastrous consequences for subalpine ecosystems. One study predicted a rapid decline with the number of trees falling to fewer than a hundred by 2150. The International Union for the Conservation of Nature, a global environmental network that tracks an organism's status, lists it as endangered. The U.S. Fish and Wildlife Service agreed that the tree deserved protection under the Endangered Species Act, but cited a lack of funding to initiate the process.

As the blister rust epidemic spreads, biologists work to stem the tide. Monitoring programs on study plots track infection and mortality rates. Studies of genetic diversity and resistance rates search for pathogen-resistant strains. Other researchers develop, propagate, and plant rust-resistant trees. This work sustains optimism, because without genetic resistance, survival prospects look especially bleak.

In addition to the blister rust fungus, the suppression of fire by land managers during most of the twentieth century also contributed to the tree's decline. Whitebark pines respond better than Engelmann spruce or subalpine fir to slow-moving ground fires, and tend to be early-comers that do well in post-fire environments. They also grow faster where the canopy is open. But the reluctance to let fire run its course in subalpine systems allows its competitors to replace it through succession. The solution, unfortunately, is not as simple as prescribing or managing high country burns. Human development extends up valleys and canyons that lead to whitebark communities and the risks of property damage or loss of lives sits squarely opposite the benefits of fire.

The native mountain pine beetle poses yet another threat to the tree. This insect army has ravaged huge tracts of forest from the Rockies north to Alaska, an area 50 times larger than that affected by forest fires, with five times the economic impact. An outbreak in British Columbia has destroyed lodgepole pine forests equivalent to two-thirds the area of Washington state. Pine beetles don't pose an immediate threat at Mount Rainier, but their preference for warmer temperatures opens the possibility that climate change will give them a foothold in the park.

The whitebark's continued decline could spell trouble for organisms that depend upon it, namely Clark's nutcrackers, squirrels, chipmunks, bears, and others. In addition, landscape-scale changes resulting from the wholesale loss of the pines in subalpine systems will reduce ecological and structural diversity. The remaining trees will become more vulnerable to forest-replacing fires or losses from insects or disease. The disappearance of this hardy, pioneering species will diminish the chances of others that colonize burned-over areas. Without the tree holding loose, rocky soils together, erosion will accelerate on steep slopes. Patterns of snow accumulation and melt out will change dramatically, affecting the timing and levels of water moving to the lowlands. The cascading effects of the loss of this subalpine tree could be far-reaching and ecologically calamitous. Although its remote location and limited value to foresters may slow it from receiving the attention its plight deserves, the whitebark pine is at the center of an evolving story in a complex web dictated by time, natural processes, and human intervention.

BEYOND THE TREES

It's June and summer elsewhere, but this landscape resembles late winter, turning slowly toward spring. On this day of incomparable beauty at the base of Burroughs Mountain, I'm soaking up the sun. Sunbathing is the perfect way to celebrate the day and the solitude, but it will be short-lived. This spot receives twice as much ultraviolet radiation than at sea level, and high-altitude sunburns are a hazard. The ever-present wind dries and nags. Normally on a Sunday afternoon, dozens of hikers would picnic here or stream past, but a heavy snowpack has kept the road to Sunrise closed for a few more days. Hiking up the south side of the ridge was the only way to get here, and finding a route was tricky because the trail lay hidden under a mantle of snow. My hiking partner and I followed two sets of week-old boot prints and animal spoor in the rotting snow. A steaming pile of bear scat glistened in the morning sun. Fresh mountain goat prints, their hoof tips splayed neatly outward, seemed to chide us for a late start. The pleasant morning chorus of newly arrived summer residents—hermit thrush, black-throated gray warbler, and Townsend's warbler—joined the resident Pacific wrens, gray jays, and varied thrushes.

It's just the two of us at Frozen Lake today, and my friend's youthful exuberance drew him to the challenge of wading through thigh-deep snow toward the Mount Fremont Lookout. I stay behind with my notebook. A Cascade golden-mantled ground squirrel panhandles shamelessly at my feet, hoping for an errant raisin or peanut. Recently emerged from hibernation, it seems genuinely excited to encounter its first hiker in eight or nine months. In spite of its endearing begging, I scold it and shoo it away.

At the edge of my peripheral vision, some movement catches my attention. I look up from my field notes to spot an American pipit, one of the handful of breeding birds found at this elevation. Pipits migrate to subalpine and alpine areas for the short summer season to breed and raise a clutch of three to seven young. They occur throughout North America and northeast Asia, wintering mostly in the southern United States and Mexico. Some birds take advantage of the region's mild maritime climate and winter near the Salish Sea.

Birds that breed up here benefit from relatively few predators and plenty of insects to feed their young. If a female, this bird has already built its cup-shaped nest of grasses and sedges on the ground in a sheltered, snow-free, rocky area. Once while hiking out to Mount Fremont, something at trail's edge startled me. It was a pipit, which I'd flushed from its nest. It lit for a moment but took flight again just over the rise and out of sight. Investigating carefully, I peeled back a palm-sized flap of turf to find five grayish-white eggs, smaller than my pinky knuckle, nestled in a shallow depression. As surprised as the pipit, I quickly replaced the sod and moved away.

Today, I watch a bird scavenge insects off the snow. These beetles, wasps, and flies have crash-landed here, carried from below on updrafts of wind. Known as arthropod fallout, they're important food items for alpine predators. Other pipit prey includes ballooning spiders, which use an unusual and extraordinary dispersal mechanism. The spider climbs to a prominence, releases silk into the air and then parachutes skyward. It controls its direction by reining in its silken sail or by re-ballooning, but prevailing winds determine the final destination. One study found 23 species of these odd, floating arachnids marooned on the Muir Snowfield.

American pipits, horned larks, and gray-crowned rosy finches forage on both snow-free and snow-covered areas, often drawn to the red or pink patches of snow algae known as watermelon snow. Also found in shades of yellow, orange, and even gray, it was first described by alpine travelers in Europe 2,000 years ago. Only through advances in microscopy have its secrets been discovered. Living in the water between snow crystals, algae, bacteria, and fungi have adapted to both long, inactive cold periods in winter and to summer's piercing sunlight and high temperatures. The pink and red colors come from pigments found inside the cells of green algae. A wide variety of snow algae lives in North America, Europe, and Antarctica, and probably evolved from soil or aquatic algae during the ice ages. Plant-like, these algae contain chlorophyll and photosynthesize. They also possess animal traits, namely an eyespot and the ability to swim. This unusual assortment of characteristics places them taxonomically in the Kingdom Protista, home of oddball organisms that can't be classified as plant, animal, or fungus.

Snow algae's complex life cycle features up to seven phases with the active ones occurring during spring and summer snowmelt. Starting at the snow–soil interface, resting spores germinate and then use their whip-like tails, or flagella, to swim in the meltwater toward the top of the snow-pack. Once there, they reproduce asexually and develop large, visible populations within a few days. They photosynthesize even at sub-freezing temperatures, reaching peak activity between 50° and 68° Fahrenheit. Cell pigments work like sunglasses to offer protection from extreme ultraviolet radiation. Thick inner and outer cell walls reduce water loss. After several more life cycle changes, the resting spores return to the snow–soil intersection where they become dormant. Eventually the trickling snowmelt once again stimulates germination and the process begins anew. Some resting spores remain viable for 25 years or longer.

The snow algae food web begins with the production of green algae. Microscopic grazers called protozoa and rotifers eat the food reserves stored within algal cells. Snow worms—also called ice worms—live in watermelon snow too, thriving until the temperature rises above 40°. Ice worms are close relatives of common earthworms, first studied on the Mount Rainier snowfields a hundred years ago. These dark reddish-brown to black, pencil-lead thick, inch-long worms feed on the algae. Spiders, springtails, stoneflies, and beetles also find nourishment there, in turn becoming food for pipits and other predators.

Many backcountry travelers believe that eating the pinkish-to-red snow causes diarrhea, but people in one study each ate over a pound of a watermelon snow "slushy" with no ill effects. Considering the protozoan and insect detritus that collects on the surface, those adventurous enough to try it should at least consider scraping off the top layer before sampling an alpine "snow cone."

Worldwide, the alpine zone—like the one at Frozen Lake that extends up and out to Third Burroughs Mountain—occupies only about 3 percent of Earth's surface and hosts just 4 percent of its plant life. The difference between it and the subalpine expanse near the Sunrise Visitor Center is the difference between treeless and treed areas, an arctic-tending climate versus a temperate-tending one. Mount Rainier's alpine zone begins at about 6,000 feet, where lingering outposts of mountain hemlock, subalpine fir, and whitebark pine mark the upper reaches of the subalpine meadows. It extends to the vegetation limit at about 7,000 feet. Known as fell fields, these stony expanses dominate the Frozen Lake and Burroughs Mountain areas. Plants cover less than half of the ground. For most hikers in this alpine landscape, accessible only between late June and October, first impressions are of windy, wide-open spaces with little plant and animal life. Freezing temperatures prevail for half the year; winter lurks just a few months away. Soil temperatures range lower than in the meadows below and the growing season averages half as long. Many alpine plants compensate by becoming active as soon as temperatures rise above freezing and begin photosynthesizing at lower temperatures than lowland plants. Most are small, ground-hugging perennials that, instead of initiating a complete life cycle each year, add to their existing biomass. Their leaves and flowers grow fewer and smaller than subalpine plants and their flower buds often form years before blooming. This conserves the energy that annuals expend when they undertake a full reproductive cycle in a single season.

Mat plants like pussypaws adapt to alpine environments by relying on a low spreading habit that allows the wind to flow over them. A deep taproot coaxes moisture from the porous volcanic soils. Dwarf lupine, much smaller than its subalpine relative, is a common resident. It takes its name from *lupus*, the Latin word for wolf, reflecting a mistaken belief

The alpine landscape of Burroughs Mountain with its sparse, ground-hugging vegetation. *Doug Scrima*

that lupines robbed nutrients from the soil. In fact, like all legumes, they add much-needed nitrogen. One of its adaptive features is the long, silvery hairs covering it entirely, imparting a pale-green appearance. Also present on other species and whether long or short, soft or bristly, they diffuse intense sunlight and reduce water loss, which is vital in sunny, windy areas that hold little moisture during the short growing season.

Another common mat plant found here is partridgefoot, a prostrate, creeping evergreen with tiny leaves that resemble a chicken's foot. It often grows on steep slopes with Tolmie's saxifrage. Together they stabilize the nutrient-poor, pumice-rich soils.

The dwarfed, low habit of the vegetation, the seeming scarcity of animals, and the cold, windy bite on all but the warmest summer days seem to deliver a tacit message: *Visit, but don't stay.* And as we've seen through the park's extensive archaeological record, people have long visited Mount Rainier's high country.

"How Did People Get Up Here if There Were No Roads?"

An unexpected, pounding rainstorm between Chinook Pass and Yakima has left the air pleasantly fresh and damp. I've come east of Mount Rainier to answer the question posed above, once asked of an archaeologist in the park. On this Labor Day weekend, I plan to trace most of the 80-mile trail that begins in the ancestral heartlands of the Yakama people in central Washington, climbs gradually over the spine of the Cascades, and then descends onto the apron of Mount Rainier. Ancestors of present-day Yakama people went there to hunt and gather resources that were unavailable at lower elevations. The ancient trail was also a major east-west crossing, both for native people for millennia and for European American settlers and traders beginning in the mid-1800s. Some sections survive as parts of established trails while remnants hang on as bits of footpaths, or barely visible ruts of a historic wagon road. Others have withered and disappeared. Still other portions lie submerged under Rimrock Lake, a byproduct of the 1925 Tieton Dam project.

Little of the trail's location and no knowledge of its cultural and natural history would exist outside of Yakama Nation oral histories without the determined efforts of Ray Paolella, whose long-time avocation has been to locate and restore the trail. Paolella's connection to the trail began indirectly when, as a teenage boy in Washington, DC, he read newspaper accounts

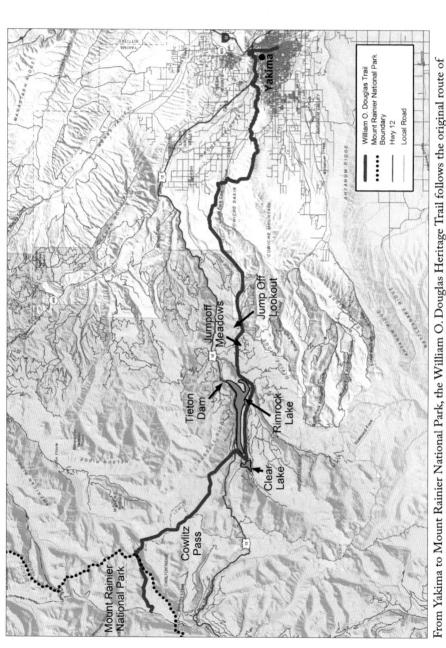

From Yakima to Mount Rainier National Park, the William O. Douglas Heritage Trail follows the original route of Native Americans, westward-bound European American pioneers, and a youthful William O. Douglas. *Yakima County GIS, modified by Kirsten Wahlquist*

of Yakima's native son, Supreme Court Justice and conservation stalwart William O. Douglas. A boyhood polio sufferer, Douglas hiked long distances and used other vigorous exercise to strengthen his legs in order to overcome his physical infirmities. His epic tales of strenuous trips into the Cascades so inspired Paolella that when he moved to Yakima years later, he searched out the trails that had helped forge Douglas's grit and fortitude. Paolella took years to piece together historic maps and other documents, interview tribal elders, and spent long days in the field reconstructing much of the trail. Before he fully understood the trail's indigenous roots stretching back over 4,000 years, Paolella's boyhood hero became the trail's namesake: The William O. Douglas Heritage Trail.

Our trek begins at Davis High School in Yakima where Douglas's life-sized bronze overlooks the courtyard of his alma mater. We pedal our mountain bikes north toward Selah Gap on paved streets and soon reach the confluence of the Yakima and Naches Rivers. Here stood *ti'mani*, a large Yakama village whose name means "pictographs made by small boy." We imagine an ancient, bustling settlement while an osprey chitters overhead.

Like the residents of other Yakama villages in this broad valley, the people of *ti'mani* lived a seasonal cycle of subsistence, gathering roots or berries, harvesting salmon, and hunting game. People celebrated the first foods in early spring. Throughout the season, salmon runs provided protein. Women dug roots in the spring, one of them the starchy, edible *Lomatium*, a carrot family member called biscuitroot or desert parsley. It played such a vital role in winter diets that some believe that it was as nutritionally important as fish and game. In summer, families moved to the cooler climes of mountain elevations. Ripe huckleberries called them to the subalpine meadows in August; they hunted there and higher up the mountainsides, too. Fall signaled their return to the valleys for more salmon runs. By mid-November, they returned to their homes to put up food for winter. Hunting and fishing continued as conditions allowed. As the daylight hours lengthened and the seasons turned, the cycle began once more.

Turning west, we soon enter Cowiche Canyon. We flush a covey of California quail. They scatter awkwardly, frantic for cover. Raucous black-billed magpies crisscross above the gravel path stretching before us as if to lead the way. The nearby village of *wa'patuxs*, which means "capital," was a gathering spot for council meetings. Another one further west was known for night fishing and hunting. As the day winds down, we load our bicycles into a waiting pickup truck and head back into town for the night.

Day two starts early and we are soon wheeling on dirt trails through the semiarid shrub-steppe landscape of central Washington. The sagebrush, perennial bunchgrasses, and broad-leaved forbs set it apart from the deserts of the southwestern United States, but rainfall here is still scarce and summers range from warm to hot. Here also are biological soil crusts comprised of mosses, lichens, fungi, algae, and cyanobacteria that create a living mulch that builds organic matter, adds nitrogen and carbon to the soil, retains moisture, and reduces erosion. The best way to study them is on hands and knees with a hand lens or through an inverted pair of binoculars. Soil crusts withstand temperature extremes and solar radiation and need little moisture to resume respiration. They're no match, however, for off-road vehicles, livestock, and even human footsteps. Easily damaged, their recovery time ranges from decades to centuries. As we move toward the Cascades, this remarkable Lilliputian community works underfoot to build and hold the precious soil.

The undulating shrub-steppe soon gives way to the ponderosa pine forests of the eastern Cascade foothills. The trail steepens and our bicycles become liabilities. In precontact times, families could have made 20 miles or more per day on foot; we struggle to cover 15. A flock of western bluebirds offers encouragement and makes for a pleasant distraction. We labor into the high country, where the guttural call of a Clark's nutcracker welcomes us. This forest reminds me of Mount Rainier's Sunrise area, but we are still two days from our destination. After more hard work and perspiration, we reach the trail's high point at Jumpoff Lookout. We gaze west into the heart of the Cascades. We'll head there tomorrow on foot, on a northwest bearing toward the mountain. Rimrock Lake sits below us, covering a once-important village called *miya'wax*. Named for a nearby high point, it meant headman or chief. A raven shows off, turning barrel rolls with pilot-like precision. We're exhausted. The heat, terrain, and distance make it difficult to follow the footsteps of early Native Americans and William O. Douglas. A friend of Ray's soon arrives in a four-wheel drive pickup to take us back to Yakima.

Shortly after sunrise the next morning, we shoulder our packs and begin hiking through the William O. Douglas Wilderness in the Okanogan-Wenatchee National Forest. The forest transitions from ponderosa pine to mostly Douglas fir, its understory rich with vanilla leaf and pipsissewa. By midday, we travel through expansive subalpine meadows brimming with bear grass and huckleberries. During a short break, we wonder whether people from *miya'wax* village made day trips here to gather grass and pick berries.

At 5,200 feet, Cowlitz Pass is one of the easiest routes through this section of the Cascades, and offers a sweeping panoramic view of Mount Rainier. The Cowlitz, Whitman, and Fryingpan Glaciers gleam in the mid-afternoon sun. Within a few hours, we'll make camp to relax and prepare for our final day's push.

After breaking camp on our last morning, I find a spot to sit quietly along the Pacific Crest Trail. A flock of yellow-rumped warblers nags for my attention, but the mountain grabs it. It dominates the western sky-line, wearing a thin, light-gray scarf of clouds that hides Little Tahoma. The Columbia Crest and Point Success summits poke out above the cloud band. From here, ancient travelers hiked either north to Chinook Pass, Tipsoo Lake, and points beyond, or continued along our route that heads nearly due west down the Laughingwater Trail to Ohanapecosh. Now on the west side of the Cascades, I see how this thread of a thoroughfare stitched people to place over great distances, binding them to resources of immeasurable value. David Rice, who as a graduate student excavated the Fryingpan Creek rock shelter, Mount Rainier's first recorded archaeolog-ical site, wrote that, "Even in prehistoric times the Cascades were more of a seasonal storehouse of goods than a barrier."

Near Ohanapecosh and the trail's end, we rest near a moss-covered nurse log while Ray sketches his vision for designation as a National His-toric Trail. Like others before it, the William O. Douglas Heritage Trail honors its travelers and their lifeways. Just as the centuries-old log hosts many plants growing upon it, this treasure of a trail holds innumerable stories about the people and the land to which they were born and bound.

STORIES IN THE GROUND

The Yakama people walking the same trail that Ray and I followed to the Ohanapecosh area might have met up with relatives or used the mineral springs there. Some continued north toward the Sunrise area to make camp on a large, relatively flat spot on the south side of Sunrise Ridge. Other people coming up the Huckleberry Creek drainage, possi-bly ancestors of Muckleshoot tribal members, might have joined them. This southwest-facing slope with fresh water and ready access to game and essential plants made for an ideal summer and early fall camp, and was used for generations. Living in temporary cedar mat or animal hide shelters, people hunted and gathered in preparation for winter. They split chores along gender lines, with the men maintaining and repairing tools, hunting, and preparing game. Women cooked, reared children, and gathered and prepared huckleberries and other plants for later use.

The berries were an indispensable overwintering staple, one of the few sources of vitamin C in early diets. People prized them for their flavor, size, and keeping abilities. They used fire to dehydrate and preserve those that they intended to use over the winter. In one method, they spread them onto mats on one side of long, low earthen mounds. Small fires near the mats dried the berries. People also laid them on mat-covered racks with small fires underneath to dry them. When dried, they poured the berries into bags or baskets and stored them in the shade where they sat until carried home.

Native Americans used fire to increase huckleberry yields. Huckleberries are one of several valuable, early successional species that grow well in recently burned-over areas. People set fires in the fall to remove unwanted trees and reduce forest cover. This created sunnier growing conditions that increased the next year's harvest of berries and other useful plants like bear grass. The influx of fresh, new growth also attracted game animals, boosting a hunter's chance of success. Rangers reported seeing these fires in the park's early days, and evidence of burns remains in erosion scars, archaeological pits, and sediment cores removed from lake bottoms. There is no way, though, to differentiate between fires set by people and those resulting from natural causes.

The rhythm of annual foraging trips to Tahoma continued for thousands of years, probably with few interruptions, until the early 1900s when the federal government officially banned native people from hunting and gathering within park boundaries. The site on the broad shoulder of Sunrise Ridge slipped out of use, living only in the memories and accounts of the people who had traveled there. The ground, too, held their stories for safekeeping.

The camp lay quietly undisturbed until 1990 when archaeologist Richard McClure Jr. found artifacts eroding out of the hillside. He dug a test pit and found more, later aged at between 450 and 2,300 years old. Reevaluation of the site later confirmed it was a singular location with such a diverse collection that park archaeologist Greg Burtchard believed it was a residential base camp used by family groups.

The park saw the value of a long-term excavation and soon joined with Central Washington University's Department of Anthropology to create a summer field school. Beginning in 1997 under the direction of anthropology professor Patrick McCutcheon, students sorted through about 180 wheelbarrow loads of material in the program's first five years. They spent thousands of hours screening soil through wire mesh screens, searching for artifacts one bucket at a time. They collected every piece of charcoal or stone tooling debris too large to slip through the screens, bagging and labeling each find for later cataloguing and analysis. Although they sampled less than one percent of the site, students recovered more than 4,300 pieces of chipped stone material, over a thousand pieces of burned bone, and found four fire hearths. An abalone shell was probably a trade item. Beads made from soft rock not found locally had been brought in from outside the area. These finds pushed the site's age to at least 4,000 years old.

I visit in August when the midday sun shows no mercy. Sunscreen, wide-brimmed hats, and plenty of water are standard equipment. Energetic and full of brio, McCutcheon spends the day teaching, directing, asking questions, and listening patiently. He takes advantage of a teachable moment to explain a difficult concept or tells a joke to lighten the mood amidst buzzing mosquitoes and perspiring brows. Later, McCutcheon explains the ultimate value of the program. While the group may recover tens of thousands of artifacts, the higher purpose extends beyond finding projectile points and stone slivers. "Our real job is connecting people with people," he says emphatically. "I want the students to think about and show respect for the people here before us and to understand the privilege of working here. Yes, I want them to get dirt under their fingernails and get bone-tired, but I want them to connect to this land and the people here before them. I believe we honor native people by doing this work respectfully, and that understanding the past is the key to anticipating the future."

McCutcheon and his students want to answer questions that will tell them how long people have come to the site, what they did here, and how their use of tools changed over time. Many answers lie embedded in the layered soils. Volcanic events over the ages deposited at least seven ash layers in this area, each with a unique signature. Similar to the Ohanapecosh Campground excavation and others on the mountain, knowing the approximate age of each layer makes it possible to date the artifacts. In the Sunrise area, the topmost tephra layer, Mount St. Helens W, resulted from rock and lava sent skyward 450 years ago. It lays on top of an orange-tinted pumice layer called Mount Rainier C that dates to around 2,300 years ago. That eruption helped rebuild the mountain to its present height following the Osceola Mudflow. Its distinctive color, identifiable even by novices like me, is an important stratigraphic marker that helps age the artifacts. Those recovered just below it, for example, are at least 2,300 years old. Digging deeper, more strata mark earlier outbursts of Mount Rainier and Mount St. Helens. A third volcano, Oregon's Mount Mazama, dropped the material that lies about three feet below ground level. Mount Rainier R, dating to about 9,000 years ago, is the oldest eruptive event captured in the ground here.

The field school is midway through its five-week session and the site hums with activity. Students on hands and knees carefully brush away soil from a grouping of fire-charred rocks a foot below ground level. They bag and label the charred earth and charcoal for later analysis that may reveal traces of hazelnuts, huckleberry seeds, or whitebark pine seeds. A student brings McCutcheon a softball-sized rock called a core, a toolmaker's starting point. He produces a small rock wedge that completes the core's original shape. A skilled knapper could fracture the wedge to make a projectile point in about 20 minutes.

Other students crouch in plots about a yard square and a yard deep, using mason's trowels and hand brooms to remove the soil. They make detailed notes and sketches of stratigraphy and of artifacts found in place. They hand buckets of soil, roots, and rock up and out of the plots for screening and close examination. Most of their finds are fingernail-sized stone chips of chert or jasper—an iron-rich, reddish chert. Once items are verified by McCutcheon or his teaching assistant, the students bag and label them without much fanfare. The work is equal parts demanding, tedious, and fascinating. Occasionally an extraordinary find punctuates the work and commands the group's attention. "Wow," a student exclaims, "look at this!" We stop our work and gather round. It's a complete, perfect projectile point, thin and symmetrical. McCutcheon calls it a Rabbit Island stemmed point, between 2,000 and 4,000 years old, named for the style first found on an island in the Columbia River. With a quiet reverence for those here long ago, we pass it around the circle in hushed voices, our questions whispered. Holding the point in my hand in the heat of the day, I feel a chill and goose bumps on my arms and neck.

On another day's visit to the site, McCutcheon led me to a towering Douglas fir near the back edge of the camp. He estimated it to be at least 500 years old; I measured it to be over 20 feet in circumference. He pointed out a faint, worn track passing in front of it that led upslope. "People probably used this trail to get to the hunting areas up above," he said. "Imagine if this tree could talk, the stories it would tell."

Some weeks later I set off with a hiking companion from the tree's base to follow the path to the high country. As it traversed the shoulder of Sunrise Ridge, we easily found some sections of trail. In other spots, it was nearly impossible to follow. We worked for hours, piecing together a route that gradually gained elevation over several miles. Eventually it leveled out and we emerged onto the broad expanse once known as Yakima Park, today more popularly known as Sunrise. The mountain dominated the southern skyline. This trail, another story in the ground, gave hunters access to some of their most highly sought-after game.

High Country Critters

In this land above the trees, hunters searched for animals not found lower down the mountain. The mountain goat, the largest mammal at this elevation, probably crossed the Bering Land Bridge some 40,000 years ago. Facing competition from the forerunners of moose and elk, it retreated to the mountains where steep slopes and scarce food presented challenges but fewer rivals.

Mountain goats frequent the Frozen Lake, Burroughs Mountain, and Fremont Lookout areas, where I often see them grazing or loafing. Looking down from Burroughs Mountain onto the tundra-like fell fields west of Frozen Lake, I once counted 47 in a loose flock. As hikers approached, the goats ambled nonchalantly into the forested area on the western flank of Mount Fremont. The five kids among them signaled that this was a group of newborns, yearlings, and two-year-olds led by parent females and others working as nannies.

Mountain goats spend nearly three-fourths of their lives on slopes steeper than 40 degrees, but they also use habitats that include forests, subalpine parklands, tundra, rocks, and snow. They move up and down the mountain according to the seasons, summering between 5,800 and 6,800 feet and wintering from 5,900 feet to as low as 4,000 feet in elevation. They use lower sites in winter to stay warmer and find food. One March afternoon, on the Wonderland Trail below Ipsut Pass in the park's northwest corner, my hiking partner made a scouting trip up ahead. He returned a short time later, breathless and excited that he had stumbled upon a mountain goat bedded down mid-trail below 4,000 feet.

Besides the Frozen Lake area, mountain goats occurred historically at over two dozen locations in the park. They frequented Cougar Rock, a scant two miles above Longmire, until the late 1920s. They are often sighted near

Skyscraper Mountain and above Summerland toward Panhandle Gap. On a recent hike of the Wonderland Trail, I camped at Summerland. Under a waxing, rising moon, I watched a kid gingerly follow close behind an adult as they crossed the rocky slopes just opposite the historic group shelter. As they traveled beyond the angle of repose, I pictured their hooves, each one a two-toed friction pad that provides superior traction on most surfaces. These eight contact points and the goat's low center of gravity add stability as it navigates yawning chasms and needle-shaped spires. One step at a time, leap after leap, the kid and five adults scrambled up impossibly rocky ledges while heading for their evening bedding-down spot. Hikers also find mountain goats around Indian Henry's Hunting Ground in the park's southwest section and on their namesake, Mount Wow. This rocky peak's name is the anglicized form of *wau*, a Sahaptin word for mountain goat.

A dozen mountain goats were introduced onto the Olympic Peninsula in the 1920s, their numbers ballooning to over a thousand within 60 years. They trampled endemic plants and eroded soils, triggering a removal program in the late 1980s that reduced their numbers to about 400. After a visitor was fatally gored in 2010, the National Park Service decided to remove them entirely from Olympic National Park. With dart tranquilizers and gun-fired nets, biologists began relocating them to isolated locations in the North Cascades, where they are a native species and less likely to damage subalpine ecosystems. The translocated mountain goats should thrive there as they increase genetic diversity and bolster sagging numbers.

The steep terrain where mountain goats spend most of their lives serves them well. They tend to be reclusive and generally don't allow humans to approach closely. When threatened or alarmed, their speed and agility enables them to move to rocky, steep terrain. Mountain goats behave differently from related species in one significant way. Their stiletto-like horns are mounted on skulls too thin for butting, so mountain goats use a ritualized fighting display that emphasizes intimidation over serious combat. Mountain goats of all ages will fiercely defend their personal space, but physical contact is usually limited to the flanks and rump, where skin up to nearly an inch thick prevents mortal injury. The skin on its backside is so impenetrable that Alaskan natives used it as chest armor for protection in battle.

Combative behavior peaks during the rutting season, the period when adults seek and acquire mates. Between fall and early winter, the males undertake the delicate process of approaching the now-aggressive females. Only the dominant animals breed, usually beginning at four or five years of age. Six months later a kid, or occasionally twins, will be born. The new-borns walk within a few hours and wean at four weeks. Parent females and other nannies are protective of the young, walking on the downhill side to guard against falls for most of the kid's first year. Outside of breeding season, males form bachelor groups or roam solo.

Winter arrives with sudden and deadly force for mountain goats, which do not hibernate. Instead, they live off stored body fat while they paw and scrape a meager diet from lichens and any available plant material. Young goats and the infirm are the most vulnerable. The slow-growing young, with a higher surface area-to-mass ratio and less insulation, lose more body heat faster and sometimes freeze to death. Besides moving to lower and warmer elevations where more food is available, physiological adaptations also help mountain goats survive the brutal cold. Their shaggy outer fur that grows up to eight inches long is the first line of defense against frigid, windy conditions. This warm, water repellent fur made mountain goats highly prized by indigenous people for clothing and blankets.

Beneath the outer fur, two to three inches of cashmere-like, dense, fine wool is the next protection against the bitter cold. Thick hides lined with body fat lie underneath. Leg hairs and hooves help conserve heat, too. Internal adaptations also play a role. The mountain goat's cardiovascular system operates at higher heart and respiration rates than humans, keeping core body temperatures elevated. Their circulatory system delivers a bare minimum of blood to lower limbs, and the ruminant digestive system derives heat from bacterial fermentation.

High country dwellers cope with the challenges of winter in a variety of ways. The American pika collects and dries grasses during the summer that it stores and eats later. Its dense, soft fur insulates it from the cold; its burrows amid rocky talus slopes protects it from most predators. The wolverine tirelessly roams huge expanses in search of food, routinely climbing sheer rock faces and avalanche chutes too dangerous for the most daring mountaineers. Last seen at Mount Rainier in 1933, the wolverine's powerful jaws crush bones left behind by other predators. These backcountry badasses will even face off against a grizzly to maintain possession of a deer carcass. On the other end of the spectrum is the hoary marmot. Members of the squirrel family, marmots weather the long, cold stretch differently from mountain goats, pikas, and wolverines: they sleep through it. They not only hibernate for up to seven months, but minimal activity marks even their wakeful periods. On a typical August day, marmots spend 12 percent of their time play-fighting, digging, and burrow inspecting, 20 percent lying about and sunning, 40 percent feeding, and the remaining time underground in their burrow. Despite spending 80 percent of their lives in their burrows, the marmot's tendency to lounge about the rocky slopes and subalpine vegetation, coupled with its photogenic personality, make it the visitor favorite at Sunrise, Summerland, and Paradise between June and September. Scan the landscape for these lovable lumps or listen carefully for their alarm calls that warn of intruders. Also known as whistle pigs because of their calls, marmots don't actually whistle. Instead, they use their vocal cords to shriek warnings when agitated or sensing danger. They're highly sociable, appear to seek each other's company, and often feed together, during which they look for and warn of predators, including coyotes, red-tailed hawks, and golden eagles. Other social behaviors include greeting each other by sniffing, nose-to-nose contact, and chewing on each other's face, neck, and back. They occasionally have boxing matches, a hilarious sight.

Marmot colonies are comprised of a reproductive male, several adult females, a few non-productive adult males, yearlings, and young of the year. Females become pregnant just after their emergence from hibernation in late May. They bear young every other year, a strategy that probably helps them restore energy reserves. Born in late June, the young emerge from the burrow a month later. With a lengthy hibernation period just ahead, putting on body fat is serious business. Fat marmots stand a better chance of surviving the winter than lean ones, and starvation during dormancy poses a greater threat than summertime predators. Like mountain goats and many other animals, the young are more susceptible to starvation than healthy adults. For the rest of the summer, the mostly herbivorous marmots eat the flowers and flower heads of paintbrush, avalanche lily, western anemone, lupine, and hellebore. Researchers at Olympic National Park found greater plant diversity in grazed than in ungrazed areas, leading them to conclude that marmots help maintain plant diversity in subalpine parklands.

By September, most marmots have the necessary body fat to carry them through the winter, so they begin spending more time resting and less time foraging. It's counterintuitive, but marmots that eat very little just before hibernating lie dormant longer than those that are heavy eaters in September. In other words, they do best by avoiding heavy snacking just prior to the long winter night.

In the account of his 1857 summit attempt, Lieutenant Kautz was the first to write of Mount Rainier's hoary marmots. "The moment anyone stirred from camp," he wrote, "a sound between a whistle and a scream would break unexpectedly…and immediately all the animals that were in sight would vanish in the earth." Hazard Stevens described marmot meat as, "extremely muscular and tough…[with] a strong, disagreeable doggy odor," but he and his climbing partner P. B. Van Trump ate the four marmots that their guide Sluiskin prepared for them during their climb in 1870.

Native Americans valued hoary marmots for food and clothing. The animal's habit of feeding and lying about in groups for more than half of their aboveground lives must have made them easy targets for hunters, and their seasonal abundance surely placed marmots into the diets of indigenous people. Their high fat content and the presence of stone scrapers at some archaeological sites within the park suggests that people scraped fat from hides and collected it for later use. Marmot pelts, at their densest concentration of fur in late summer and early fall, probably made for plush robes and blankets. Some Native Americans, however, did not hunt hoary marmots. For at least some Yakama people, they evoked the supernatural. Those who traveled alone could fall prey to the marmot's whistling, called onward until eventually losing all sense of time, space, and self.

The Last Hunters

When park ranger Thomas O'Farrell found an abandoned Indian camp in July 1915 in Yakima Park, complete with shelters and horse corrals, he wrote to Park Supervisor Dewitt L. Reaburn. O'Farrell asked "whether or not the Indians under their treaty with the federal government have the right to hunt and take game in national parks and if not what steps are to be taken to cause a discontinuance of this practice." Similarly unsure of how to respond, Reaburn sought guidance by writing to the Secretary of the U.S. Department of the Interior in Washington, DC.

The next month, O'Farrell dispatched two assistant rangers to Yakima Park where they encountered about 30 Yakama Indians at *Me~yah~ah Pah*, "Place of the Chief," a spot where they had camped for generations. With a young woman in the group who had attended the Cushman Indian School assisting as translator, the rangers announced that it was illegal to hunt within park boundaries. Through her, the group's leader asserted that the Yakima Treaty had reserved them access to traditional hunting, fishing, and gathering places like Yakima Park. Lacking knowledge of the treaty, the two bid the group farewell and reported back to their supervisor.

Reaburn waited patiently for a response from the Interior Department while headlines in the Tacoma *Daily Ledger* proclaimed "Indians Defy Park Hunting Regulations." The matter of whether native people should hunt and gather at Mount Rainier eventually made its way to the Interior's Office of the Solicitor. When Solicitor General Preston C. West issued his opinion that fall, it shocked everyone following the case. West's interpretation was that the Mount Rainier National Park Act of 1899 did not suspend Indian treaty rights and that the federal government could not prohibit their hunting within park boundaries. He cited a long-standing principle of Indian law that ambiguities must be resolved as understood by signers at the time of the treaty. West recommended the negotiation of an agreement that would "permit the Indians to enjoy 'the privilege of hunting, gathering roots, and berries…' with such restrictions as may be necessary to prevent…the killing of game beyond the reasonable needs for subsistence."

West's view contradicted the mind-set of the times by seeing no conflict between Indian treaty rights and the mission of the national parks. In the aftermath of the removal of native people from Yosemite, Yellowstone, Glacier, and other now-federal lands, West's unexpected opinion not only went against the grain, it was unpopular. It also went unheeded. The department's decision to disregard his recommendation to uphold treaty rights at Mount Rainier sprang from an established practice of separating Native Americans from their homelands-turned-national parks.

In the meantime, judicial and legislative action transpired that would have a bearing on the ability of native people to hunt and gather on their ancestral lands that were now part of Mount Rainier National Park. A court decision established state jurisdiction over game management. Then, the Act of Congress of June 30, 1916, gave park rangers the authority to make arrests. In the fall of 1916, Supervisor Reaburn received approval to arrest Indian hunters in the park. In October of the following year, he got word of a group in Yakima Park. With a ranger and State Game Commissioner and former Park Supervisor Edward S. Hall, the trio drove all day over rough roads from Longmire to the White River area. Early the next morning they hiked up to the encampment, arrested six Yakama Indians, and escorted them back to the car for an impromptu court appearance with Commissioner Hall. The rangers confiscated the rifles of the men, who pled guilty to charges of illegal hunting. They were released with light fines.

The net effect of the arrests ended subsistence hunting within the park, an initiative that came as much from local sources given voice in the area's newspapers than from any coherent policy from the nascent National Park Service.

In his 1963 ethnographic study of the Native American tribes surrounding Mount Rainier, anthropologist Allan H. Smith noted that the people he interviewed expressed bitterness at the suspension of their traditional subsistence activities on Tahoma's slopes. Yet in much the same way that a river or a herd of mountain goats cannot be contained within human-made boundaries, the seasonal return of people was a force more powerful than any map or Act of Congress. Even after the bungled affair at Sunrise's Yakima Park, rangers encountered native people picking huckleberries in mountain meadows. Scores of other trips likely went unnoticed.

Mount Rainier National Park staff has worked diligently in recent years to heal and build relationships with the tribes near Tahoma. In the late 1990s, park staff began working with tribal councils and their cultural resource staff to collaborate on archaeological excavations and other projects. Elders told stories that filled information gaps.

One hundred years after the Yakima Park embarrassment, feelings of mutual trust continue to build. One of the strongest indicators of restored relations is the return of Native American cultural practices within park boundaries. The Nisqually Indian Tribe uses a collecting permit to harvest plants that do not occur in the Puget Sound lowlands. It also makes exclusive use of some campsites in the Longmire area adjacent to the Nisqually River. The designation provides tribal members with a specific use area and recognizes its long-term use by native people. The Cowlitz Indian Tribe has a similar set-aside area in the Ohanapecosh Campground in the park's southeast corner. These and other initiatives demonstrate the compatibility of the National Park Service mission with honoring the enduring presence of Native Americans at Mount Rainier. Like the circle of a hundred-plus archaeological sites around the mountain, all involved help build a ring of goodwill and understanding.

Conclusion:
Messengers for the Future

To celebrate the end of my fieldwork for this book, I set out on a six-day solo hike of the park's 93-mile Wonderland Trail. As I start counterclockwise from Sunrise, a red-tailed hawk soaring in tight circles overhead seems like a good omen. A sooty grouse shares a section of trail, unconcerned at my close approach.

Along the trail, rock cairns guide me through the rocky moonscape between Seattle and Spray Parks, and through a washed out crossing at the North Mowich River. At Golden Lakes at nightfall, a barred owl hoots its eternal questions: "Who cooks for you? Who cooks for you-all?" Other animals busily prepare for winter. A northwest deer mouse feeds intently at mid-trail, oblivious to my presence. Cheeks stuffed with seeds for storing, a ground squirrel charges to within 10 feet before darting for cover.

A Clark's nutcracker leads the way up Cowlitz Divide as I round north and back toward Sunrise. Scores of bird droppings mark the trail, stained blue from huckleberries. Set off the busy Wonderland Trail, the Olallie Creek camp is used so little that the gray jays have forgotten about it. Olallie is the Chinook Jargon word for berries—I pick two kinds for my morning gruel without even leaving my campsite.

On my final morning and within a few hours of closing my loop, I return to the site of the park's first archaeological excavation. This under-sized, protected camp in the Fryingpan Creek drainage sheltered small groups of hunters for over 1,500 years, and is the bedrock underlying much of our understanding of the presence of first people at Tahoma. Some years before, sitting on the natural rock bench along the shelter's back wall, I began to feel the presence of the mountain's forebears. I gained a new understanding of the mountain that reaches back over 9,000 years and will extend forward long after me. At last fully aware that I walked in the footsteps of the ancients, I became follower instead of leader, learner instead of teacher, and scribe instead of writer. I set off from the rock shelter and before long spot 12 mountain goats grazing on Goat Island Mountain. After more than a dozen trips up this valley over the last 40 years, this is the first time I've seen the mountain's namesake on its hillsides.

It's a fitting conclusion to a noteworthy hike.

Before hundreds of thousands of day hikers and 12,000 or more backpackers began using the Wonderland Trail each year, it played a critical role in resource protection. Built by park rangers between 1907 and 1911, the trail allowed them to move around the mountain in order to protect against poachers, vandals, and fire. Congressional funding in 1915 helped complete a circuit that ran longer than today's version. The Mountaineers, a Seattle- and Tacoma-area based outdoor club formed in 1906, wasted no time in pressing their boot prints upon the newly forged footpath. Packhorses carried provisions to feed the club's 90-plus hikers. Fifty-seven made the successful "side trip" from Glacier Basin to the summit. The first group to circumambulate the mountain, they took three weeks to complete their trek.

In 1920, Superintendent Roger Toll penned this grand description: "There is a trail that encircles the mountain. It is a trail that leads through primeval forest, close to the mighty glaciers, past waterfalls and dashing torrents, up over ridges and down into canyons; it leads through a veritable wonderland of beauty and grandeur." A century later, Toll's words still ring true. Weather permitting, the Wonderland Trail repeatedly offers changing, breathtaking views as hikers tramp around the imposing monolith. Prominent features like Sunset Amphitheater, Gibraltar Rock, or Willis Wall situate us in the landscape and overwhelm us with their magnificence. The 1915 Mountaineers trip reporter wrote that the mountain's changing faces caused him to "gasp and gaze in silent awe and wonder."

For anyone looking to add another dimension to their relationship with Mount Rainier, there are multiple opportunities. A constricting federal budget has left the National Park Service with a maintenance backlog of almost $12 billion. The work needed at Mount Rainier totals nearly $300 million. The funding shortfalls, the immediacy of some of the projects, and the willingness of people to roll up their sleeves has created a vibrant era of volunteerism at the park. Some visitors join one-day events like National Public Lands Day each September. Others create a legacy of service that spans decades. Whatever the commitment, the park welcomes visitors to participate in hands-on service. In a recent 10-year stretch, an annual average of nearly 1,900 volunteers—individuals, families, and service groups—contributed over 716,000 hours of labor, a value of $17.7 million. Flexible hours, fantastic scenery, and knowledgeable park employees make for a memorable experience. Some of the most popular opportunities include trail maintenance, revegetation, and the Meadow Rover program. Citizen science projects like MeadoWatch and the Cascade Butterfly Project offer opportunities to work alongside highly trained technicians collecting data that informs research. More on these and other opportunities can be found in Appendix B: Steward Mount Rainier.

As I sat at the Carbon River's edge one July morning, a female merganser and one of her young splashed down directly in front of me. These diving ducks with serrated bills patrol rivers and lakes to catch fish, amphibians, and invertebrates. The pair began working their way upriver against the onrushing torrent. Bobbing wildly and using their wings as oars, they struggled forward, fell back, and then drove ahead once more. The birds' brief display carried the message and captured the spirit of how we must steward Mount Rainier as we approach the quarter post of this century. The direct effects of climate change upon the mountain are clear. Glacial recession releases an unprecedented volume of sediment into the lowland river systems, creating puzzles for municipalities and problems for numerous organisms. No fewer than nine species occurring at Mount Rainier or in its neighboring watersheds receive protection under the U.S. Endangered Species Act: bull trout, Chinook salmon, golden paintbrush, marbled murrelet, Mazama pocket gopher, northern spotted owl, Puget Sound steelhead, streaked horned lark, and Taylor's checkerspot butterfly. Others deserve protection. While some native species decline, many invasives increase. These are but a few of the consequences of climate change.

As trustees of one of North America's most precious and alluring landscapes, our actions—or lack thereof—will shape the Mount Rainier National Park that future generations will inherit. Park staff, researchers, elected officials, and we the public must shoulder the significant challenges and enormous responsibilities that come with stewardship of this unmatched natural treasure.

On a snowy December day at Longmire, about 60 people, many of them children, gathered for a historic event. The snowfall alternated between light flurries and heavy, fat flakes. Forearm-length icicles hung like dragon fangs from the eaves of the Community Building. We circled around park wildlife ecologist Tara Chestnut and Washington Department of Fish and Wildlife biologist Jeff Lewis to learn about fishers, a member of the weasel family long absent from the North Cascades. Trappers extirpated them from the region in the early twentieth century before trapping regulations were in place and their warm, dense fur brought $300 per pelt—equivalent to over $3,500 today. Now on this wintry morning, six fishers were to be released as part of a reintroduction project. The three females and three males had been live-trapped from the southern end of their present range near Williams Lake, British Columbia, over 300 miles north of Mount Rainier. After enduring a 16-hour ordeal in individual plywood cartons in the back of a pickup truck, the animals were about to be set free in their new home. Today's delivery was the third in the North Cascades; the other animals were accounted for and doing well. Following their reentry into Washington's wilderness, researchers track them by listening for signals from each fisher's transmitter.

When the time came to set the translocated animals free, Chestnut and Lewis encouraged "the younger biologists to come forward to release the fishers." About 10 kids, ages four to 12, eagerly approached the crates containing the anxious occupants. With the help of the agency personnel, the children removed the wooden doors from one end of each box to allow the animals to become part of the first fisher population at Mount Rainier in a hundred years. Sleek and low-slung, about the size of a house cat and trailing a tail longer than its body, each animal charged through the snow and disappeared into the forest.

On a Saturday morning with 16 inches of snow on the ground, I'd just witnessed a momentous triumph. This milestone, the product of a partnership between the National Park Service, the Washington Department of Fish and Wildlife, the U.S. Fish and Wildlife Service, Conservation Northwest, and the University of Washington, brought people of all ages together to return wildlife to their original territory, exemplifying our highest endeavor for Tahoma and its people.

Acknowledgments

In 2009 I searched without success for a current natural history of Mount Rainier for a course I was preparing to teach at The Evergreen State College in Olympia, Washington. I wasn't looking for a textbook, but something similar to Tim McNulty's now-classic *Olympic National Park: A Natural History*. After our tenacious reference librarians also conceded defeat, several colleagues badgered me to undertake my first book project. Despite apprehension about properly treating such a wide berth of topics, I reluctantly agreed to take it on. The demands of a busy teaching schedule slowed my progress at times, but the unwavering interest and support of Evergreen students, staff, and faculty, friends, family, and Mount Rainier National Park staff sustained me. I am indebted to them all. Largely because of them, you are now holding the book that I imagined using with my students.

Early on, Evergreen's Bill Ransom and Sandra Yannone cajoled and eventually convinced me to accept the challenge. Soon joined by Tim McNulty and Sean Williams, this quartet of seasoned, accomplished writers mentored me throughout the entire process. Any inaccuracies or shortcomings in the text, however, are solely my responsibility.

Park scientists and other staff at Mount Rainier were integral to my fieldwork. Their standard trick was to invite me to join them on their research and after exchanging early morning salutations, ask me, "By the way, can you carry this?" Now retired, archaeologist Greg Burtchard included me in three excavations and gave generously of his time to introduce me to the fascinating story of the mountain's cultural resources. He read multiple drafts, his careful eye keeping me to known facts and avoiding speculation, and provided valuable feedback on images used in the book. National Park Service regional geomorphologist Paul Kennard and park geologist Scott Beason let me tag along on fieldwork and directed me to relevant scientific literature about the effects of climate change and aggradation at Mount Rainier. Scott dutifully read and commented on the book's geological accounts and assisted with images. Keith Bagnall allowed me to accompany his northern spotted owl survey crew and guided my research of marbled murrelets. He also read and gave feedback on those sections. I assisted Ben Wright and Scott Anderson in the field where I learned firsthand the mountain's key stories about fish and amphibians, respectively. Aquatics ecologist Rebecca Lofgren supported my work by providing access to researchers and discussing ongoing projects. Kevin Bacher shared the park's volunteer program data, and his staff coordinated my overnight stays in the Longmire area.

Other readers, many of whom are experts in their field, included Adrian Wolf, Andy Lockhart, Barbara Leigh Smith, Carolyn Driedger, Dimitri Antonelis Hunter, Donny Stevenson, George Antonelis, Jr., George Walter, Jim Gawel, Joe Kane, John McLain, Kelli Bush, Mary Linders, Michelle Aguilar-Wells, Pat McCutcheon, Nalini Nadkarni, Patrick Pringle, Paul Przybylowicz, Ray Paolella, Rich Wells, Sarah Pedersen, Sheri Tonn, and Wilma Cabanas.

Those helping with research, providing feedback, or helping in some other valuable way included Abbi Wonacott, Adam Martin, Amy Callahan, Andrea Heisel, Barbara Samora, Barry Penland, Beau Antonelis, Bill Gesler, Bill and Lois Wilson, Blake Smith, Bob and Anne Smith, Bridgid Cummins, Brooke Childrey and the Mount Rainier National Park Archives staff, Carl Elliott, Carl Fabiani, Carlos Diaz, Casey Goldin, Cindy Marchand Cecil, Courtney Purdin, Curtis D. Tanner, David McAvity, David Rice, Davy Clark, Dennis Suarez, Dorothea Collins, Doug Jones, Doug Scrima, Dylan Fischer, Emily Zazz Brouwer, Emmett O'Connell, Eric D. Forsman, Eric Walkinshaw, Evergreen student interns in the Carolyn Dobbs Mount Rainier Internship Program and students in *Mount Rainier: The Place and Its People* and related programs, Evergreen's faculty librarians and other reference librarians, Frederica Bowcutt, Hans Hunger, Jane Hall, Jay Satz, Jerry Hedlund, Jill Bushnell, Jim Lynch, Jim Vallance, Jennifer Cutler, Joe Barrentine, Joe Mabel, John C. Miles, Judy Bentley, Keith Lazelle, Ken Tabbutt, Kyle Bustad, Liza Rognas, Lydia Sigo, Marcia Zitzelman, Matt Kastberg, Maureen Ryan, Merrill Peterson, Mike Martian, Miko Francis, Mindy Muzatko and the circulation staff at Evergreen's Daniel J. Evans Library, Pam Udovich, Pat Dunn, Paul McMillin, Peggy Lovellford, Phil Hoge, Rita Rosenkranz, Roger Andraczik, Russ Ladley, Ryan Richardson, Shane Jeffries, Tara Chestnut, Tarin Todd, Walter Siegmund, and Zoltan Grossman.

Evergreen Faculty Emerita Lucia Harrison, with whom I taught courses on Mount Rainier, graciously provided the pen and ink drawings that introduce each chapter, and one additional image. Graphics artist and scientific illustrator Kirsten Wahlquist modified or created many images, providing a steady hand and a consistent touch.

Faculty at Evergreen's *Write That Book!* workshops supplied camaraderie, ideas, and inspiration. The college supported my work with Sponsored Research grants and a sabbatical. The Mazamas, a group dedicated to mountaineering education in the Portland, Oregon area, also helped with funding.

Several friends in Indian Country shared their cultural knowledge and language before I even began the project; no one foresaw the eventual significance of their contributions. I raise my hands in gratitude to my Wuhlshootseed language teachers Doris Allen, Zalmai Zahir, and Donna Starr. Thanks also to Sharon Hamilton and master weaver Yvonne Peterson for introducing me to the rudiments of collecting and working with cedar.

My hiking companions who accompanied me on innumerable trips into the field were extraordinarily helpful, providing all manner of physical and moral support. At all times of day or night, on overnight trips or long days and sometimes in bitter weather, they shared coffee and snacks, assisted with observations, asked excellent questions, made wonderful suggestions, and buoyed my spirits dozens of times. Simply stated, they helped me do that which I could not do alone. Abram Bell and Bob Maxwell served in this capacity early in the fieldwork. Robb Pack did so in the latter stages. Dimitri Antonelis Hunter and Valerie Antonelis-Lapp were indispensable from start to finish.

The staff at Washington State University Press was extremely helpful and patient, especially with this first-time author. Special thanks to my editor Beth DeWeese, whose "less is more" style helped me get to the point countless times.

And finally, some who dearly loved the mountain and encouraged my efforts did not live to see the book in print. I salute the memories of Carolyn Dobbs, Rob Smurr, and Russ Gibbs.

To all these people—and to any I may have inadvertently omitted—I offer my heartfelt thanks for your contributions to *Tahoma and Its People*. I hope that your connection to Mount Rainier is well represented in this book.

Appendix A: Explore Mount Rainier

The nineteenth century English biologist Thomas H. Huxley wrote, "To a person uninstructed in natural history, his country or seaside stroll is a walk through a gallery filled with wonderful works of art, nine-tenths of which have their faces turned to the wall." With ample preparation, every Mount Rainier outing brings new and exciting discoveries, making its "wonderful works of art" available to all of us.

In addition to the Ten Essentials—map, compass, sunglasses and sunscreen, extra clothing, headlamp or flashlight, first-aid supplies, firestarter, matches, knife, and extra food—binoculars and field guides are indispensable for getting the most out of every trip. Many outdoor enthusiasts consider their camera gear essential; cellphone cameras suffice for the rest of us. Field guides and mobile identification apps are abundant—research and select those that best meet your needs. My favorite all-in-one guide is *Natural History of Pacific Northwest Mountains* by Dan Mathews.

Wildlife

Seeing wildlife is largely a matter of being in the right place at the right time, but there are ways to increase your odds. Most animals are highly sensitive to human scent and sounds, so travel in small groups as quietly as possible. Scan the landscape frequently for movement or irregularities. That lump in the meadow may be a hoary marmot, and that mass on a limb a barred owl. Get an early start to avoid crowds of people that may inadvertently drive animals away. Many animals are most active in the early daylight hours and at dusk. Seasons matter: you'll find black bears more easily during August and September in subalpine meadows than in winter, when they're hibernating. Winter, on the other hand, is an ideal time to study tracks in the snow or in riverside mud or sand bars. My favorite reference is Dave Moskowitz's comprehensive and well-illustrated *Wildlife of the Pacific Northwest*.

Remember that these are wild animals, and that interacting with them is not only strictly prohibited, it's dangerous for both you and the animal. Feeding them or being careless with food and garbage can make them food-conditioned, associating humans with food. Even if it's the cutest chipmunk, squirrel, or rare Cascade red fox panhandling for your snack, don't feed it! Shoo it away and keep your food out of reach. The park's *Keep Wildlife Wild* program teaches people that refraining from feeding wildlife is one of the best ways to help them. See a ranger for details and buy a nifty button that supports the program.

In terms of where to find wildlife, remember these tips and know how to respond in the event of encounters. At any picnic area, campground, pullout, or popular trailside spot, expect uninvited guests. Facial stripes differentiate chipmunks from squirrels, and both may show up at snack time. Gray jays or the stunning, black-crested Steller's jay appear out of nowhere, looking for handouts. In any of these areas, you can find Columbia black-tailed deer, abundant from the lowland forests to the subalpine meadows. Also look along roadsides and at forest edges.

Elk are larger but more elusive, frequenting lower elevations in winter and moving to higher ground in summer. Find their large, hamburger bun-shaped tracks in mud or sand. Listen in the fall for their eerie bugling, thought to aid in herding cow elk.

Raccoons are common, especially around water, but are stealthy and more active at night. Coyotes are similarly secretive, but look for their prints mid-trail. With practice, you can note the star-shaped pattern in the negative space between their toes and footpad.

On rocky hillsides, listen for the familiar "Eek!" of the American pika, which sounds much like a squeeze toy. You may find these tiny rabbit relatives or their over-wintering hay piles among the talus slopes between Sunrise and Frozen Lake, in the Paradise trail system, or in similar habitat throughout the park.

In the subalpine meadows, the crowd-pleasing show-stealer is the photogenic hoary marmot. These rodents have a short active season of June through September but their aboveground feeding and loafing habits make them easy to find, especially early in the day. Look for movement in the meadows, find them sunning on rocks, and listen for their whistle-like alarm calls.

Another abundant subalpine mammal is the mountain goat, often seen lounging or grazing on the alpine tundra west of Frozen Lake and in the Burroughs Mountain, Skyscraper Mountain, and Fremont Lookout areas. Another reliable spot is above Summerland toward Panhandle Gap. If you find them, keep your distance and back away if they hold their ground.

Mount Rainier's high country hosts several regional specialty birds that are highly sought after by birdwatchers. Visitors seeking white-tailed ptarmigan, gray-crowned rosy finch, or boreal owl should consult *A Birder's Guide to Washington*.

Count yourself among the lucky few if you run across black bear, cougar, fisher, marten, mountain beaver, northern flying squirrel, opossum, porcupine, skunk, or weasel. In all cases, maintain a safe distance. Cougar and bear merit special precautions.

Known by more common names than any other North American mammal (mountain lion, puma, catamount, and Florida panther), cougars are rare but present at Mount Rainier. They prefer to avoid people but if you meet one, stay calm and stay put. Don't turn your back and don't run. Grab small children and group up with your companions to appear larger. If the cougar moves toward you, move your arms and make a ruckus. Face the animal as you back away slowly. If it attacks, fight back aggressively. Use your trekking poles as weapons or throw dirt in its eyes.

There are few records of black bear attacks on people in the United States, and none at Mount Rainier. As with all other wild things, don't feed or attract bears through carelessness with food. If you encounter them, announce your presence loudly and continue to make noise. Look around to see that you are not between a sow and its cubs. Step up onto a nearby rock to appear larger, to better survey the situation, and to find an exit route. Most charges are bluffs, so stand your ground. If attacked, fight back. In the worst-case scenario, curl up in a ball and play dead. With these precautions, I enjoy safely searching for black bears in early spring in the Stevens Canyon road pullouts. Later in the season, bears can pop up in any spot containing ripe huckleberries, like Berkeley Park, Shadow Lake, and around Frozen Lake on the park's north side. Food-conditioned "problem bears" sometimes occur, and are an entirely different story. Take seriously any warning signs at backcountry camps or campgrounds and take extra precautions. Finally, report any cougar or black bear sightings to park staff.

WILDFLOWERS

Mount Rainier's wildflower spectacular can be phenomenal, but remember that the beginning of each season depends on the depth of the previous winter's snowpack and the rate at which it melts out in the spring and early summer. This makes it difficult to predict the bloom's peak, although it generally falls between late July and early August. Check with the visitor centers or the park's trail conditions web page that usually includes status reports. Paradise attracts tens of thousands of visitors each summer, and weekends can be especially crowded, so visit mid-week and arrive early if you can. If you must go on weekends, arrive and depart early. Brace yourself for jammed parking lots, lengthy restroom lines, and a seemingly endless wave of people.

If you want to savor wildflowers without the crowds, and you enjoy hiking, the upper reaches of the Eagle Peak trail boasts acres of flowers, as does Spray Park, accessed from Mowich Lake in the park's northwest corner. Many hikes in the Sunrise area meander through subalpine meadows, and the Shriner Peak trail on the east side also has large expanses above the low-lying old growth forest. The Wonderland Trail winds through lots of flower-laden parklands, too.

GLACIERS AND RIVERS

Getting up close to Mount Rainier's glaciers and rivers enables one to appreciate the power and might of these master landscape shapers and water carriers, but demands the utmost caution. Most people are unprepared for glacier travel and should never step foot onto them. The same advice holds for the rivers, where frigid, forceful currents can sweep even the strongest hikers off their feet. Do not attempt to wade into, swim in, or cross them.

In spite of these dangers, several hikes and road pullouts allow close access to glacial termini and their rivers. At Paradise, the Nisqually Vista Trail offers excellent views of the glacier and the Moraine Trail a look at its terminus. Near Cougar Rock Campground, the Carter Falls pullout provides safe access to the Nisqually River. Further downriver at Longmire behind the residential complex, it runs along an access road that fronts the Longmire Community Building. Look carefully for evidence of erosion resulting from high water events. This area is ideal for those with accessibility needs.

The 17-mile round trip to the Carbon Glacier terminus lies outside the range of many day hikers, but mountain bikes can make the adventure more manageable. Ride the five miles from the road's gate to Ipsut Creek camp, lock your bicycles at the end of the road, and hike the three-plus miles to find a lunch spot near the glacier's snout.

To approach the White River, drive State Route 410 from Enumclaw or west from State Route 123 at Cayuse Pass, looking for pullouts along the river. A wide parking area at milepost 59 has good views and some space for exploring; notice the height of the riverbed relative to the road's elevation.

At Sunrise, the Emmons Vista and other nearby trails lend good views of the Emmons Glacier terminus, headwaters of the White River.

The Ohanapecosh River runs clear on the park's east side, carrying surface and groundwater instead of glacial runoff. The Grove of the Patriarchs Trail parallels and then crosses it via a suspension bridge.

For more accessibility information about these and other features, visit the park website at www.nps.gov/mora.

Appendix B: Steward Mount Rainier

Whether you're a Puget Sound local who treats Mount Rainier as your extended backyard or one of the thousands of annual national or international visitors, anyone can expand and deepen their relationship with the mountain. One way is to participate in ranger-led hikes or interpretive programs. Or, use a trail guidebook to explore new areas or stay overnight at any of the two inns, four campgrounds, or 40-plus wilderness trailside camps. Volunteer for part of a day or a long-term project, as an individual or as part of a family or other group. Visit the park's Volunteer Program Homepage at www.nps.gov/mora/getinvolved/volunteer.htm. Read the Volunteer FAQs, and learn more about opportunities like the following.

Trail Work and Related Projects

The park keeps an updated, online listing of volunteer projects—most are available from May through September. The season ends on National Public Lands Day, a park-wide workday on the fourth Saturday in September. Other organizations, including those listed below, lead volunteer work at Mount Rainier:

- Mount Rainier National Park Associates, mrnpa.org, organizes a handful of work projects each summer. They welcome newcomers.
- The Student Conservation Association (SCA), www.thesca.org, has put thousands of young people to work on public lands—and transformed their lives—for over 60 years. Crews work at Mount Rainier most summers and in all 50 states. Twelve percent of all National Park Service staff are SCA alumni.
- Northwest Youth Corps, www.nwyouthcorps.org, patterned after the Civilian Conservation Corps of the 1930s and 1940s, began offering education-based work experiences to young people in 1984.
- Washington Trails Association, www.wta.org, a resource to hikers since the 1960s, has organized and led trail maintenance projects for over 25 years. It offers volunteer opportunities statewide, including within and around Mount Rainier, from one-day work parties to eight-day "Volunteer Vacations."

To learn more about volunteering, visit their websites. Similar steward-ship opportunities exist at most national parks and other federal agencies, listed at www.volunteer.gov.

CITIZEN SCIENCE

There are multiple approaches to citizen science, most centered on the general public working alongside scientists and technicians to collect and analyze data. The park conducts amphibian and butterfly surveys (the latter as part of the Cascade Butterfly Project). Both programs collect data that helps scientists understand the effects of climate change on these sensitive organisms.

MeadoWatch, www.meadowatch.org, sponsored by the University of Washington, also focuses on the effects of climate change at Mount Rainier. Researchers train citizen scientists to make observations and record data relating to the flowering sequence of the park's subalpine wildflowers.

The Dragonfly Mercury Project, www.nps.gov/articles/dragonfly-mercury-project.htm, is a newer addition to the citizen science offerings. Experts work with volunteers to collect dragonfly larvae, which can live up to nine years. Their lives as insect-eating predators places them relatively high in aquatic food chains where they bio-accumulate mercury, a toxic pollutant.

OTHER WAYS TO MAKE A DIFFERENCE

With needed maintenance at our national parks totaling nearly $12 billion, there's never been a better time to help steward them. If on-the-mountain volunteer work is impractical, there are other ways to make an impact. Email or write the U.S. Department of the Interior, which administers the National Parks. Or write your congressional representatives—especially after a visit—to encourage them to sponsor legislation that resolves the funding shortfalls.

Washington's National Park Fund, wnpf.org, is the official fundraiser for the state's three largest parks: North Cascades, Olympic, and Mount Rainier, which benefit from the sales of national park license plates and the generosity of donors. Local residents wishing to connect with their watershed have several opportunities, including the Puyallup Watershed Initiative, www.pwi.org, and the Nisqually River Council, nisquallyriver.org.

Appendix C: Significant Geologic Events at Mount Rainier

What Happened?	How Many Years Ago?	What Time on 24-Hour Clock?
A series of explosive eruptions begin to create Mount Rainier	Between 500,000 and 420,000	From midnight to 5 a.m.
Fewer and smaller eruptions continue	Between 400,000 and 280,000	From 4:48 to 10:34 a.m.
A lava flow creates Rampart Ridge near Longmire	370,000	6:14 a.m.
More mountain-building lava flows	280,000-190,000	From 10:34 a.m. to 2:53 p.m.
Lava flows create Little Tahoma Peak	195,000-130,000	From 2:38 to 5:46 p.m.
Alpine glaciers proliferate and extend down river valleys for many miles—the Hayden Creek Glaciation	170,000-130,000	From 3:50 to 5:46 p.m.
Lava flows create Echo and Observation Rocks above Spray Park	120,000	6:14 p.m.
A lava flow creates Mazama Ridge near Paradise	90,000	7:41 p.m.
Lava flows create the upper mountain	40,000-20,000	From 10 to 11 p.m.
Another period of alpine glaciers—the Evans Creek Glaciation	22,000-15,000	From 10:57 to 11:17 p.m.
One of the world's largest lahars—the Osceola Mudflow	5,600	11:44 p.m.

Continued

What Happened?	How Many Years Ago?	What Time on 24-Hour Clock?
More eruptions, lahars, and lava flows—the Summerland Period	2,700-2,200	From 11:52 to 11:54 p.m.
The most recent major lahar, the Electron Mudflow	500	11:58:30 p.m.
Steam clouds, ashfall, and pumice	200-125	From 11:59:27 to 11:59:40 p.m.
Glaciers have receded the furthest up mountain since record keeping began in 1905	Present day	11:59:59 p.m.

Notes

Introduction

Names of the mountain as used by Native Americans: Allan H. Smith, *Takhoma: Ethnography of Mount Rainier National Park* (Pullman: Washington State University Press, 2006); Cecelia Svinth Carpenter, *Where the Waters Begin: The Traditional Nisqually Indian History of Mount Rainier* (Seattle: Northwest Interpretive Association, 1994); and Judy Wright and Carol Ann Hawk, *History of the Puyallup Tribe of Indians* (Tacoma, WA: Puyallup Tribal Council, 1997).

1. The Place: Geologic History and Processes

Ancestor Mountains

Crandell's notebook: U.S. Geological Survey research geologist Jim Vallance shared a PDF of Rocky Crandell's field notes from July 25, 1953. The notes showed Crandell's "light bulb" and ideas about the deposits that later became known as the Osceola Mudflow.

Plate tectonics and other basic geological principles: Stephen L. Harris, *Fire Mountains of the West: The Cascade and Mono Lake Volcanoes* (Missoula, MT: Mountain Press, 1988); Robert J. Lillie, *Parks and Plates: The Geology of Our National Parks, Monuments and Seashores* (New York: W. W. Norton, 2005); Brian J. Skinner et al., *Dynamic Earth: An Introduction to Physical Geology*, 5th ed. (Hoboken, NJ: Wiley Custom Services, 2008); Hill Williams, *The Restless Northwest: A Geological Story* (Pullman: Washington State University Press, 2002).

Signs of Mount Rainier as a living volcano: Patrick T. Pringle et al., *Roadside Geology of Mount Rainier National Park and Vicinity* (Olympia, WA: Washington Division of Geology and Earth Resources Information Circular 107, 2008); "Volcano Seismicity: Mount Rainier," Pacific Northwest Seismic Network, pnsn.org/volcanoes/mount-rainier#description.

Geological formations: Pringle, *Roadside Geology*.

Creative Forces, Destructive Powers

Geologic history of Mount Rainier: Dwight R. Crandell, *The Geologic Story of Mount Rainier*, Geologic Survey Bulletin 1292 (Washington, DC: U.S. Government Printing Office, 1969); Richard S. Fiske et al., *Geology of Mount Rainier National Park Washington*, Geologic Survey Professional Paper 444 (Washington, DC: U.S. Government Printing Office, 1963). The most recent account of the mountain's geology is Pringle's *Roadside Geology*, which provides full detail on events on the 24-hour clock and a more complete record of Mount Rainier's geology.

Cordilleran Ice Sheet: R. M. Thorson, "Ice-sheet Glaciation of the Puget Lowland, Washington, during the Vashon Stade (Late Pleistocene)," *Quaternary Research* 13 (1980): 303–21.

Recent eruptions of Mount Rainier: Pringle, *Roadside Geology*, details eruptions during the nineteenth century, including those covered by the *Seattle Post-Intelligencer* and the *Tacoma Daily Ledger* in 1894. Also see Harris, *Fire Mountains of the West*, and *Fire and Ice: The Cascade Volcanoes* (Seattle: Mountaineers, 1976).

Estimations of property values: Recep Cakir and Timothy J. Walsh, *Loss Estimation Pilot Project for Lahar Hazards from Mount Rainier, Washington*. Washington Division of Geology and Earth Resources, Information Circular 113 (Olympia: Washington State Department of Natural Resources, 2012).

Warning Signs of Climate Change

Glacier basics: Carolyn L. Driedger, *A Visitor's Guide to Mount Rainier Glaciers* (Longmire, WA: Pacific Northwest Parks and Forest Association, 1986).

Predictions of glacial recession worldwide: Intergovernmental Panel on Climate Change, *Climate Change 2007: Impacts, Adaptation and Vulnerability* (Cambridge: Cambridge University Press, 2007).

Glacial recession at Mount Rainier: Jon Riedel and Michael A. Larrabee, *Mount Rainier National Park Glacier Mass Balance Monitoring Annual Report, Water Year 2011* (Fort Collins, CO: National Park Service 2015); Scott R. Beason, *Change in Glacial Extent at Mount Rainier National Park from 1896 to 2015* (Fort Collins, CO: National Park Service, 2017).

Loss of eight glaciers at Mount Rainier: National Park Service geomorphologist Paul Kennard, Glacier Research and Geohazards Workshop, February 19, 2013; personal communication, November 15, 2017.

Paradise ice caves: W. R. Halliday and C. H. Anderson Jr., *The Paradise Ice Caves, Mount Rainier National Park* (n.p., 1972); Mauri Pelto, "Paradise Glacier Ice Caves Lost," glacierchange.wordpress.com/2010/04/29/paradise-glacier-ice-caves-lost.

From Mountain Heights to Salish Seas

Nisqually Glacier stagnant ice: Scott Beason and Paul Kennard, personal communications, August 25 and September 8, 2011; Laura Walkup and Justin Ohlschlager, "Nisqually Glacier Mapping and Stagnant Ice Survey Field Protocol" (unpublished report, 2012).

Glacier outburst floods: Carolyn L. Driedger and Andrew G. Fountain, "Glacier Outburst Floods at Mount Rainier, Washington State, USA," in *Annals of Glaciology* 13 (1989), 51–55. I accompanied Paul Kennard and four other geologists on August 1, 2017, on a one-day trip to the near-terminus of the South Tahoma Glacier, headwaters for Tahoma Creek.

Imminent Threats Program: Robert P. Jost, Jeni Chan, and Paul Kennard, "The Westside Road: Physical Flood Protection in Mount Rainier National Park, Washington," in Beason et al., *Rivers Gone Wild: Extreme Landscape Response to Climate-Induced Flooding and Debris Flows* (Seattle: National Park Service, 2017). Field Trip Guide to Tahoma Creek and other sites of glacier outburst floods as part of the Geological Society of America Annual Meeting on October 21, 2017.

Lahars: James W. Vallance, "Lahars," and Kelvin S. Rodolfo, "Notes from the Hazard from Lahars and Jökulhlaups," in *Encyclopedia of Volcanoes* (San Diego: Academic Press, 2000); James W. Vallance and Kevin M. Scott, "The Osceola Mudflow from Mount Rainier," *Geological Society of America Bulletin* 109:2 (1997), 143–63; Pringle, *Roadside Geology*; Dwight R. Crandell, *Postglacial Lahars from Mount Rainier Volcano, Washington*, Geological Survey Professional Paper 67 (Washington, DC: U.S. Government Printing Office, 1971); David A. John et al., "Characteristics, Extent, and Origin of Hydrothermal Alteration at Mount Rainier Volcano," *Journal of Volcanology and Geothermal Research* 175 (2008): 289–314.

Aggradation at Mount Rainier: Scott R. Beason, "The Environmental Implications of Aggradation in Major Braided Rivers at Mount Rainier National Park, Washington," (master's thesis, University of Northern Iowa, 2007); Scott R. Beason and Paul M. Kennard, "Environmental and Ecological Implications of Aggradation in Braided Rivers at Mount Rainier National Park," *Natural Resource Year in Review—2006* (National Park Service, 2007); Scott R. Beason, Laura C. Walkup, and Paul M. Kennard, *Aggradation of Glacially-sourced Braided Rivers at Mount Rainier National Park, Washington: Summary Report for 1997–2012* (National Park Service, 2014); and Jonathan A. Czuba et al., *Geomorphic Analysis of the River Response to Sedimentation Downstream of Mount Rainier, Washington*, U.S. Geological Survey Open-File Report 2012-1242, 2012.

Presidentially declared flood disasters and county flood management plans: *2013 Flood Hazard Management Plan Update, King County, Washington; Pierce County Rivers Flood Hazard Management Plan, Vol. 1,* Pierce County Public Works, 2013.

2. The People: Footprints of Days Past

People on the Landscape

Ohanapecosh Campground excavation: Greg Burtchard invited me to three archaeological excavations in the park between 2011 and 2015. Field notes from the Ohanapecosh Campground sites, September 24 and 30, 2015.

Salish Sea Neighbors

Nineteenth century diseases in Indian communities: Robert T. Boyd, *The Coming of the Spirit of Pestilence: Introduced Infectious Diseases and Population Decline Among Northwest Coast Indians, 1774-1874* (Seattle: University of Washington Press, 1999).

Languages in the Salish Sea region: Wayne Suttles and Barbara Lane, "Southern Coast Salish" in *Handbook of North American Indians*, 1998; Smith, *Takhoma*.

Chinook Jargon: George Lang, *Making Wawa: The Genesis of Chinook Jargon* (Vancouver: University of British Columbia Press, 2008); George Coombs Shaw, *The Chinook Jargon and How to Use It* (Seattle: Rainier Printing Co., 1909); Edward Harper Thomas, *Chinook: A History and Dictionary of the Northwest Coast Trade Jargon* (Portland, OR: Binford and Mort, 1935).

First Salmon Ceremonies: Erna Gunther, "An Analysis of the First Salmon Ceremony," *American Anthropologist* 28 (1926), 605–17; Erna Gunther, "A Further Analysis of the First Salmon Ceremony," *University of Washington Publications in Anthropology* 2, no. 5 (1928), 129–73. Each spring, the Muckleshoot Indian Tribe holds a first salmon ceremony and dinner.

Longhouses: Suttles and Lane, "Southern Coast Salish"; Hilary Stewart, *Cedar: Tree of Life of the Northwest Coast Indians* (Seattle: University of Washington Press, 1984); Barry M. Pritzker, *Native Americans: An Encyclopedia of History, Culture and Peoples*, vol. 1 (Santa Barbara: ABC-CLIO, 1998).

Arborvitae, the Tree of Life

Importance and uses of red cedar: Stewart, *Cedar*; Erna Gunther, *Ethnobotany of Western Washington*, rev. ed. (Seattle: University of Washington Press, 1973); Jim Pojar and Andy MacKinnon, *Plants of the Pacific Northwest Coast* (Vancouver: BC Ministry of Forests/Lone Pine Publishing, 2004).

Account of a recent tribal canoe journey: Will Bill Jr., "Canoe Journey to Lummi 2019 is Now History," *Muckleshoot Messenger* 20, no. 5, August 8, 2019. www.muckleshoot.nsn.us/media/49275/muckleshoot messenger august 2019r.pdf.

The Myth: "Indians Were Very Superstitious and Afraid of It"

Nisqually demarcation line around the mountain: Carpenter, *Where the Waters Begin*.

"Young Man's Ascent of Mount Rainier": Arthur C. Ballard et al., *Mythology of Southern Puget Sound: Legends Shared by Tribal Elders* (North Bend, WA: Snoqualmie Valley Historical Museum, 1999).

Alison Brown and Yakama hunters: Aubrey L. Haines, *Mountain Fever: Historic Conquests of Rainier* (Seattle: University of Washington Press, 1999); Dee Molenaar, *The Challenge of Rainier* (Seattle: Mountaineers, 1971); Edmond Meany, *Mount Rainier: A Record of Exploration* (New York: Macmillan Company, 1916).

Early Native American trails: Smith, *Takhoma*; multiple conversations with Greg Burtchard, 2011–2015.

Tolmie's trip to the mountain: William F. Tolmie, *The Journals of William Fraser Tolmie, Physician and Fur Trader* (Vancouver, CA: Mitchell Press Ltd., 1963). Haines, *Mountain Fever*; Meany, *Mount Rainier*.

Sluskin and the two surveyors: Lucullus V. McWhorter, "Chief Sluskin's True Narrative," *Washington Historical Quarterly* 8, no. 2 (1917): 96–101; Haines, *Mountain Fever*; Michael F. Turek and Robert H. Keller Jr., "Sluskin: Yakima Guide to Mount Rainier," *Columbia* 5, no. 1 (Spring 1991): 2–7.

Kautz summit attempt: August V. Kautz, "Ascent of Mount Rainier," *The Overland Monthly* 14, no. 5 (May 1875), 393–403; Gary Fuller Reese, *Nothing Worthy of Note Transpired Today: The Northwest Journals of August V. Kautz*, Tacoma, WA: Tacoma Public Library, 1978; "First Attempted Ascent, 1857," in Meany, *Mount Rainier*; Haines, *Mountain Fever*; Molenaar, *Challenge of Rainier*.

Stevens and Van Trump ascent: "First Successful Ascent, 1870," Meany, *Mount Rainier*; Molenaar, *Challenge of Rainier*; Haines, *Mountain Fever*.

Assumptions about Native Americans avoiding mountainous areas: *Smith*, Takhoma.

Emerging Truth, Stubborn Bias

Smith and Daugherty research: Smith, *Takhoma*.

Fryingpan Creek Rock Shelter: David G. Rice, *Archaeological Test Excavations in Fryingpan Rockshelter, Mount Rainier National Park* (Pullman: Washington State University, 1965); Patrick M. Lubinski and Greg C. Burtchard, "Archaeology of Fryingpan Rockshelter (45PI43) in Mount Rainier National Park," *Archaeology in Washington* 11 (2005): 35–52.

Two excavations near Sunrise: Greg C. Burtchard, Stephen C. Hamilton, and Richard H. McClure, Jr., *Environment, Prehistory and Archaeology of Mount Rainier National Park, Washington* (Seattle: National Park Service, 1998, repr. 2003); Eric A. Bergland, *Archaeological Test Excavations at the Berkeley Rockshelter Site (45-PI-303), Mt. Rainier National Park, Washington*, SRD Report of Investigations No. 1, 1988.

Burtchard's archaeological models for Mount Rainier: Burtchard et al., *Environment, Prehistory and Archaeology*.

3. The Nisqually River, From Glacier to Sound

Subalpine Meadows and the Mountain Hemlock Zone

Naming of Paradise: There are differing accounts of to whom the "Paradise" quote is attributed. Virinda Longmire is named in Gary Fuller Reese, *Mount Rainier National Park Place Names* (Tacoma, WA: Tacoma Public Library, 2009) and at www.historylink.org/File/10711; her daughter-in-law Martha Longmire is attributed at www.nps.gov/mora/planyourvisit/paradise.htm.

John Muir's climb of Mount Rainier: William F. Badé, *The Life and Letters of John Muir* (Boston: Houghton Mifflin, 1924).

Paradise as a wildflower wonderland: Bob Gibbons, *Wildflower Wonders: The 50 Best Wildflower Sites in the World* (Princeton, NJ: Princeton University Press, 2011).

Calliope and Rufous hummingbirds: John Martin, "Different Feeding Strategies of Two Sympatric Hummingbird Species," *The Condor* 90 (1988): 233–36; Lazarus Walter Macior, "The Pollination Ecology of Pedicularis on Mount Rainier," *American Journal of Botany* 60:9 (1988), 863–71.

Phenology studies at Paradise: Elli J. Theobald et al., "Climate Drives Phenological Reassembly of a Mountain Wildflower Community," *Ecology* 98 (2017), 11.

Anna Wilson et al., "Monitoring Wildflower Phenology Using Traditional Science, Citizen Science and Crowdsourcing Approaches," *Park Science*, 33:1 (2016).

Mission of the National Park Service: www.nps.gov/grba/learn/management/organic-act-of-1916.htm.

Impacts and recovery time at higher elevations: Regina Rochefort and Darrin Swinney, "Human Impact Surveys in Mount Rainier National Park: Past, Present, and Future," *USDA Forest Service Proceedings* RMRS-P-15-VOL-5, 2000; Ann H. Zwinger and Beatrice E. Willard, *Land Above the Trees: A Guide to American Alpine Tundra* (New York: Harper and Row, 1972).

People of the Grass

Nisqually legend of how they came to the region: Carpenter, *Where the Waters Begin*.

Lushootseed words in the text: Lushootseed is traditionally a spoken, not a written, language. Linguists and tribal elders developed an alphabet and orthography in the twentieth century. Few fluent speakers of Lushootseed remain, but most local tribes have language programs. Lushootseed words included here seek to honor the land's first people and their language.

Leschi: Ezra Meeker, *The Tragedy of Leschi* (Seattle: Historical Society of Seattle and King County, 1980).

Qu-Lash-qud's paradise: Charles Wilkinson, *Messages from Frank's Landing: A Story of Salmon, Treaties and the Indian Way* (Seattle: University of Washington Press, 2000).

Prairies Found and Lost

Effects of undercooked camas: Gary E. Moulton, ed., *The Journals of the Lewis and Clark Expedition*, vol. 6 (Lincoln: University of Nebraska Press, 1983); Jack Nisbet, *The Collector: David Douglas and the Natural History of the Northwest* (Seattle: Sasquatch Books, 2009).

Changes to Puget Sound prairies: Sarah T. Hamman et al., "Fire as a Restoration Tool in Pacific Northwest Prairies and Oak Woodlands: Challenges, Successes and Future Directions." *Northwest Science* 85:2 (2011), 317–28.

Threatened Species, Drastic Measures

Streaked horned lark: Derek W. Stinson, *Periodic Status Review for the Streaked Horned Lark in Washington* (Olympia: Washington Department of Fish and Wildlife, 2016); Washington Department of Fish and Wildlife, *Threatened and Endangered Wildlife in Washington: 2012 Annual Report*; Adrian Wolf, personal email communications, December 2018.

Use of prescribed fire on prairies: Hamman, "Fire as a Restoration Tool"; Erik J. Rook et al., "Responses of Prairie Vegetation to Fire, Herbicide and Invasive Species Legacy," *Northwest Science* 85:2 (2011), 288–302.

Flowing forward

Trova Heffernan, *Where the Salmon Run: The Life and Legacy of Billy Frank Jr.* (Seattle: University of Washington Press, 2012).

J. Michael Reed, Lewis W. Oring, and Elizabeth M. Gray. "Spotted Sandpiper (*Actitis macularius*)," Birds of North America Online (2013), *doi.org/10.2173/bna.289*.

4. Historic Longmire and the Surrounding Area

Land of Disturbance Events

Longmire lahars: Pringle, *Roadside Geology*; Crandell, *Postglacial Lahars*.

Nature's Wetlands Engineer

American beaver natural history and economic value: Dietland Müller-Schwarze and Lixing Sun, *The Beaver: Natural History of a Wetlands Engineer* (Ithaca, NY: Cornell University Press, 2003); Moulton, *Journals of Lewis and Clark*.

Hudson's Bay Company: Müller-Schwarze and Sun, *Beaver*.

Parachuting beavers: Elmo W. Heter, "Transplanting Beavers by Airplane and Parachute," *Journal of Wildlife Management*, 14:2, 1950.

Beaver reservoirs and firefighting: Müller-Schwarze and Sun, *Beaver*.

Nature's Woodland Excavator

Pileated woodpecker natural history: Evelyn L. Bull and Jerome A. Jackson, "Pileated Woodpecker (*Dryocopus pileatus*)," Birds of North America Online (2011), birdsna.org. *doi.org/10.2173/bna.148*.

The Soul of the Forest

Spotted owl: R. J. Gutiérrez, A. B. Franklin, and W. S. Lahaye, "Spotted Owl (*Strix occidentalis*)," Birds of North America Online (1995), birdsna.org. *doi.org/10.2173/bna.508*.

Barred owl range expansion: Kurt M. Mazur and Paul C. James, "Barred Owl (*Strix varia*)," *Birds of North America Online* (2000), birdsna.org. doi.org/10.2173/bna.148.

Spotted owl population in Canada: Jared Hobbs, "Habitat Action Plan for Northern Spotted Owl Pursuant to the Species at Risk Act" (2019).

5. The Puyallup River: Watershed under Pressure

The River's End

Wilkes's description of Commencement Bay and the people living there: Charles Wilkes, *Narrative of the United States Exploring Expedition, during the years 1838, 1839, 1840, 1841, 1842, vol. II.* (London: Ingram, Cooke, and Co., 1852); Francis B. Barkan, *The Wilkes Expedition: Puget Sound and the Oregon Country* (Olympia: Washington State Capital Museum, 1987); Blumenthal, R. W., *Charles Wilkes and the Exploration of Inland Washington Waters: Journals from the Expedition of 1841* (Jefferson, NC: McFarland and Co., 2009).

First People, First Stewards

Meanings of the word "Puyallup": Dawn E. Bates, Thom Hess, and Vi Hilbert, *Lushootseed Dictionary* (Seattle: University of Washington Press, 1994); www.puyallup-tribe.com/ourtribe.

Ethnography of Puyallup villages and lifeways: Marian W. Smith, *The Puyallup-Nisqually* (New York: Columbia University Press, 1940).

The Bay: Developing, Degrading, and Detoxifying

Development, decline, and restoration of Commencement Bay: Murray Morgan, *Puget's Sound: A Narrative of Early Tacoma and the Southern Sound* (Seattle: University of Washington Press, 2003); Hart Crowser, Sitcum/Blair/Milwaukee Waterway Nearshore Fill (2015), www.hartcrowser.com/sitcum.html; U.S. Environmental Protection Agency Region 10, *Fourth Five-Year Review Report for Commencement Bay* (Seattle, 2014), www3.epa.gov/region10/pdf/sites/cb-nt/4th_fyr_cbnt_2014.pdf; Washington State Department of Ecology (DOE), "Toxics Cleanup in Commencement Bay"; Washington State DOE, Urban Waters Initiative (2015), City of Tacoma, WA, The Foss Waterway Cleanup, www.cityoftacoma.org/government/city_departments/environmentalservices/surface_water/restoration_and_monitoring/thea_foss_waterway_cleanup.

Tacoma's stormwater and water quality of Commencement Bay: Sheri Tonn interview, November 17, 2015.

More Troubled Waters

Population growth in the Puyallup basin: Puyallup River Watershed Council, *Puyallup River Watershed Assessment* (unpublished draft, 2014).

Ratings of Puyallup watershed waterways and Pierce County Surface Water Report Cards: Pierce County Public Works and Utilities Surface Water Management Division. *2013 Report Card: Surface Water Health*, www.co.pierce.wa.us/ArchiveCenter/ViewFile/Item/2612; Washington State DOE, *Water Quality Assessment and 303(d) List* (2015). www.ecy.wa.gov/programs/Wq/303d/index.html; Washington DOE, Water Quality Improvement Projects—Puyallup River Basin Area Projects (2015). www.ecy.wa.gov/programs/wq/tmdl/puyallup/index.html.

Fight Like a Salmon

Bull trout life history and field surveys: Shelley A. Spalding, "Bull Trout: Big Beautiful Fish in Small Pristine Streams" (master's thesis, The Evergreen State College, 1994); D. B. Stewart et al., "Fish Life History and Habitat Use in the Northwest Territories: Bull Trout (*Salvelinus confluentus*)" (Ottawa: Fisheries and Oceans Canada, 2007); field notes written while accompanying Mount Rainier fisheries biologist Ben Wright, September 17, 2015.

Decline of salmon populations: National Oceanic and Atmospheric Administration (NOAA) Fisheries, (2015). *Chinook Salmon (Oncorhynchos tshawytscha)*. www.fisheries.noaa.gov/species/ chinook-salmon-protected; Jim Lichatowich, *Salmon Without Rivers: A History of the Pacific Salmon Crisis* (Washington, DC: Island Press, 1999); Jim Lichatowich, *Salmon, People and Place: A Biologist's Search for Salmon Recovery* (Corvallis: Oregon State University Press, 2013).

Puget Sound Chinook Salmon Recovery Plan: Millie M. Judge, "2011 Implementation Status Assessment. Final Report. A Qualitative Assessment of Implementation of the Puget Sound Chinook Recovery Plan," Seattle: National Marine Fisheries Services.

Call Them Kings

Benefits of salmon carcasses on terrestrial environments: C. Jeff Cederholm et al., "Pacific Salmon and Wildlife: Ecological Contexts, Relationships, and Implications for Management," in *Wildlife-Habitat Relationships in Oregon and Washington*, David H. Johnson and Thomas A. O'Neill, eds. (Corvallis: Oregon State University Press, 2001); James M. Helfield, and Robert J. Naiman. "Effects of Salmon-Derived Nitrogen on Riparian Forest Growth and Implications for Stream Productivity," *Ecology* 82, no. 9 (Sept. 2001): 2403–9.

A Mosaic in Disrepair

Native American stories about salmon: Vi Hilbert, ed., *Haboo: Native American Stories from Puget Sound* (Seattle: University of Washington Press, 1985); Arthur C. Ballard, *Mythology of Southern Puget Sound: Legends Shared by Tribal Elders* (North Bend, WA: Snoqualmie Valley Historical Museum, 1999); Lichatowich, *Salmon Without Rivers*.

New White River fish passage trap and haul operation: U.S. Army Corps of Engineers, Seattle District, Final Environmental Assessment for Mud Mountain Dam Upstream Fish Passage, Pierce County, Washington. www.nws.usace.army.mil/Portals/27/docs/environmental/resources/ 2015EnvironmentalDocuments/Draft_FINAL_Fish_Trap_and_Barrier_Structure_EA_4-30-15_ FINAL.pdf?ver=2015-05-05-125719-650.

Great Mountain, Grave Danger

Aging the Electron Mudflow: Patrick T. Pringle, "Buried and Submerged Forests of Oregon and Washington: Time Capsules of Environmental and Geologic History," *Western Forester*, 59:2 (2014), 14–15, 22; Pringle, *Roadside Geology*, 191.

Detecting hydrothermal alteration: David A. John et al., "Characteristics, Extent and Origin of Hydrothermal Alteration at Mount Rainier Volcano, Cascades Arc, USA," *Journal of Volcanology and Geothermal Research* 175, no. 3 (2008): 289–314. doi:10.1016/j.jvolgeores.2008.04.004; Carol A. Finn, Thomas W. Sisson, and Marla Deszcz-Pan. "Aerogeophysical Measurements of Collapse-Prone Hydrothermally Altered Zones at Mount Rainier Volcano," *Nature* 409 (2001): 600–603.

6. The Carbon River Area: Land of Moisture

Damp Valley, Lush Forest

Uses of devil's club as medicine: Gunther, *Ethnobotany of Western Washington*; Pojar and MacKinnon, *Plants of the Pacific Northwest Coast*.

Protection afforded by thorny or prickly plants: Ernest Small and Paul M. Catling, *Canadian Medicinal Crops* (Ottawa: NRC Research Press, 1999).

Columbine and native children: Gunther, *Ethnobotany of Western Washington*.

Wonders Large and Small

Lichens in old growth forests Bruce McCune et al., "Epiphyte Habitats in an Old Conifer Forest in Western Washington, USA," *The Bryologist* 103:3 (2000), 417–27.

Canopy studies: Nalini Nadkarni, *Between Earth and Sky: Our Intimate Connection to Trees* (Berkeley: University of California Press, 2008); Adrian Lance Wolf, "Bird Use of Epiphyte Resources in an Old-Growth Coniferous Forest of the Pacific Northwest" (master's thesis, The Evergreen State College, 2009); Camila Tejo Haristoy et al., "Canopy Soils of Sitka Spruce and Bigleaf Maple in the Queets River Watershed, Washington," *Soil Science Society of America Journal* 78, no. S1 (2014).

Nadkarni canopy root research: Nalini M. Nadkarni, "Canopy Roots: Convergent Evolution in Rainforest Nutrient Cycles," *Science* 214:4524 (1981), 1023–4; Nalini M. Nadkarni, "Roots That Go Out on a Limb," *Natural History* 94:5 (1985), 42–49; Haristoy et al., "Canopy Soils."

Riddles of the Fogbird

Tlingit story about murrelet and raven: Frederica De Laguna, *Under Mount Saint Elias: the History and Culture of the Yakutat Tlingit* (Washington, DC: Smithsonian Institution Press, 1972).

The search for murrelet nest sites: Harry R. Carter and Spencer G. Sealy, "Who Solved the Mystery of the Marbled Murrelet?," *Northwestern Naturalist* 86 (2005): 2–11.

Murrelet nesting status and population trends: Gary A. Falxa and Martin G. Raphael, "Northwest Forest Plan—The First Twenty Years (1994–2013)," USDA, Forest Service, Pacific Northwest Research Station, 2016.

"Marbled Murrelet (*Brachyramphus marmoratus*)," Birds of North America Online, doi:10.2173/bna.276.

And the People Came

Artifacts recovered in the Carbon River watershed: Burtchard et al., *Environment, Prehistory and Archaeology.*

The Around-the-Mountain Road: C. Cain and S. Dolan, *Cultural Landscape Inventory, Carbon River Road, Mount Rainier National Park* (unpublished draft, 2006), National Park Service. Ashford, WA: Mount Rainier National Park.

Civilian Conservation Corps at Mount Rainier: Theodore Catton, *Wonderland: An Administrative History of Mount Rainier National Park* (Seattle: National Park Service Cultural Resources Program, 1996); Cain and Dolan, *Cultural Landscape Inventory.*

Long-term alpine lakes study: Field interviews with Scott Anderson, Rebecca Lofgren, and Ben Wright, August 9 and August 15, 2016.

7. The Sunrise Area: The High and Dry East Side

Of Nuts and Nutcrackers

Clark's nutcracker life history: Diana F. Tomback, "Clark's Nutcracker (*Nucifraga columbiana*)," Birds of North America Online (1998), doi:10.2173/bna.331.

Trouble in the High Country

Whitebark pine mortality: Regina M. Rochefort, "The Influence of White Pine Blister Rust (*Cronartium ribicola*) on Whitebark Pine (*Pinus albicaulis*) in Mount Rainier National Park and North Cascades National Park Service Complex, Washington," *Natural Areas Journal*, 28:3 (2008), 290–98.

Prediction of whitebark pine decline: Gregory J. Ettl and Nicholas Cottone, "Whitebark Pine (*Pinus albicaulis*) in Mt. Rainier National Park, Washington, USA: Response to Blister Rust Infection." In *Species Conservation and Management: Case Studies*, 36–47. H. Resit Akçayaya et al., eds. (Oxford: Oxford University Press, 2004).

Protection needed under Endangered Species Act: Craig Welch, "Rainier's Trees Could Hold Key to Saving Whitebark Pines," *Seattle Times*, July 22, 2011.

Beyond the Trees

Ballooning spiders: Crawford, Rodney L. and John S. Edwards, "Ballooning Spiders as a Component of Arthropod Fallout on Snowfields of Mount Rainier, Washington." *USA Arctic and Alpine Research*, 18:4 (1986), 429–37.

Study of people eating watermelon snow: David C. Fiore, Denise D. McKee, and Mark A. Janiga. "Red Snow: Is it Safe to Eat? A Pilot Study," *Wilderness and Environmental Medicine*, 8 (1997), 94–95.

"How Did People Get Up Here if There Were No Roads?"

Lomatium in native diets: Eugene S. Hunn and David H. French, "Lomatium: A Key Resource for Columbia Plateau Native Subsistence," *Northwest Science* 55:2 (1981), 87–94.

The Cascades as a storehouse of goods: David G. Rice, "Indian Utilization of the Cascade Mountain Range in South Central Washington," *Washington Archaeologist*, 8:1 (1964).

High Country Critters

Marmots: Tim McNulty, *Olympic National Park: A Natural History* (Seattle: University of Washington Press, 2009); Kautz, "Ascent of Mount Rainier"; Edmond S. Meany, *Mount Rainier*; Eugene S. Hunn and James Selam, *Nch'i-Wána "The Big River": Mid-Columbia Indians and Their Land* (Seattle: University of Washington Press, 1990).

Bibliography

Abbe, Tim, George Pess, David R. Montgomery, and Kevin L. Fetherston. "Integrating Engineered Log Jam Technology into River Rehabilitation." In *Restoration of Puget Sound Rivers*, David R. Montgomery, Susan Bolton, Derek B. Booth, and Leslie Wall, eds. Seattle: University of Washington Press, 2003.

ABR, Inc. Radar Observations of Marbled Murrelets in Mt. Rainier National Park, Washington, Unpublished, 2002.

Acker, Steven A., Jerry F. Franklin, Sarah E. Greene, Ted B. Thomas, Robert Van Pelt, and Kenneth J. Bible. "Two Decades of Stability and Change in Old-Growth Forest at Mount Rainier National Park." *Northwest Science*, 80:1 (2006), 65–72.

Allaback, Sarah and Victoria Jacobson. *100 Years at Longmire Village*. Seattle: Northwest Interpretive Association, 1999.

Allaby, Ailsa and Michael Allaby, eds. *A Dictionary of Earth Sciences*, 2nd ed. Oxford: Oxford University Press, 1999.

Altman, Bob. "Historical and Current Distribution and Populations of Bird Species in Prairie-Oak Habitats in the Pacific Northwest." *Northwest Science,* 85:2, 194–222, 2011.

Andrews, James. "Glaciers." In *Vital Signs Monitoring: Overviews*. The North Coast and Cascades Science Learning Network and the Inventory and Monitoring Program. U.S. Department of the Interior, National Park Service, 2011.

Andrews, Rebecca W. "Hiaqua: Use of Dentalium Shells by the Native Peoples of the Pacific Northwest." Master's thesis, University of Washington, 1989.

Anonymous. "Memorial from the Geological Society of America Favoring the Establishment of a National Park in the State of Washington." In *The Miscellaneous Documents of the Senate of the United States for the Second Session of the Fifty-Third Congress.* Washington, DC: Government Printing Office, 1895.

Antor, Ramón J. "The Importance of Arthropod Fallout on Snow Patches for the Foraging of High-Alpine Birds." *Journal of Avian Biology*, 26:1 (1995), 81–85.

Apostol, Dean and Marcia Sinclair. *Restoring the Pacific Northwest: The Art and Science of Ecological Restoration in Cascadia*. Washington, DC: Island Press, 2006.

Arno, Stephen F. and Carl E. Fiedler. *Mimicking Nature's Fire: Restoring Fire-Prone Forests in the West*. Washington, DC: Island Press, 2005.

Associated Press. NBC News.com. (2006). *Flood Damage at Mount Rainier Extensive*. November 13, 2006. www.nbcnews.com/id/15651997/ns/weather/t/flood-damage-mount-rainier-extensive/#. U7TLPk13vtQ.

Aubrey, Dennis A., Nalini M. Nadkarni, and Casey P. Broderick. "Patterns of Moisture and Temperature in Canopy and Terrestrial Soils in a Temperate Rainforest, Washington." *Botany* 91 (2013): 739–744.

Aubry, Keith B. and Catherine M. Raley. *The Pileated Woodpecker as a Keystone Habitat Modifier in the Pacific Northwest*. U.S. Department of Agriculture, Forest Service, Gen. Tech. Rep. PSW-GTR-181, Berkeley, CA, 2002a.

———. "Selection of Nest and Roost Trees by Pileated Woodpeckers in Coastal Forests of Washington." *Journal of Wildlife Management* 66:2 (2002b), 392–406.

Bachelet, Dominique, Bart R. Johnson, Scott D. Bridgham, Pat V. Dunn, Hannah E. Anderson, and Brendan M. Rogers. "Climate Change Impacts on Western Pacific Northwest Prairies and Savannas." *Northwest Science* 85:2 (2011), 411–29.

Badé, William F. *The Life and Letters of John Muir*. Boston: Houghton Mifflin, 1924.

Ballard, Arthur C., Kenneth G. Watson, and Snoqualmie Valley Historical Museum. *Mythology of Southern Puget Sound: Legends Shared by Tribal Elders*. Reprint of the 1929 Edition. North Bend, WA: Snoqualmie Valley Historical Museum, 1999.

Balston, Carolyn L. *Timber, Trains and Tides: A Brief History of Tacoma, Washington*. Tacoma: Pacific Lutheran University, 1988.

Banta, Robert M. "The Role of Mountain Flows in Making Clouds." *Atmospheric Processes over Complex Terrain, Meteorological Monographs*, 23:45 (1990), 229–282.

Barash, David P. "Ecology of Paternal Behavior in the Hoary Marmot (*Marmota caligata*): An Evolutionary Interpretation." *Journal of Mammalogy*, 56 (1975): 613-618.

———. *Marmots: Social Behavior and Ecology*. Stanford, CA: Stanford University Press, 1989.

———. "Pre-Hibernation Behavior of Free-Living Hoary Marmots, *Marmota caligata*." *Journal of Mammalogy*, 57 (1976): 182-185.

———. "The Social Behaviour of the Hoary Marmot (*Marmota caligata*)." *Animal Behaviour* 22 (1974): 256–61.

Barkan, Frances B. *The Wilkes Expedition: Puget Sound and the Oregon Country*. Olympia: Washington State Capital Museum, 1987.

Bates, Dawn E., Thom Hess, and Vi Hilbert. *Lushootseed Dictionary*. Seattle: University of Washington Press, 1994.

Beason, Scott R. *Change in Glacial Extent at Mount Rainier National Park from 1896 to 2015*. Natural Resource Report NPS/MORA/NRR—2017/1472 (Fort Collins, CO: National Park Service, 2017).

———. "The Environmental Implications of Aggradation in Major Braided Rivers at Mount Rainier National Park, Washington." Master's thesis, University of Northern Iowa, 2007.

Beason, Scott R., Laura C. Walkup, and Paul M. Kennard. *Aggradation of Glacially-Sourced Braided Rivers at Mount Rainier National Park, Washington: Summary Report for 1997–2012*. Natural Resource Technical Report NPS/MORA/NRTR—2014/910. Fort Collins, CO: National Park Service, 2014.

Beason, Scott R., and Paul M. Kennard. "Environmental and Ecological Implications of Aggradation in Braided Rivers at Mount Rainier National Park." In *Natural Resource Year in Review—2006, A Portrait of the Year in Natural Resource Stewardship and Science in the National Park System*. Denver: National Park Service, 2007.

Beason, Scott R., Paul M. Kennard, Christopher S. Magirl, Joseph L. George, Robert P. Jost, and Taylor R. Kenyon. *Rivers Gone Wild: Extreme Landscape Response to Climate-Induced Flooding and Debris Flows, and Implications to Long-Term Management at Mount Rainier National Park*. Field Trip Guide #410, 2017 Geological Society of America Annual Meeting. Seattle: National Park Service, Mount Rainier National Park, 2017.

Bédard, Jean. "New Records of Alcids from St. Lawrence Island, Alaska." *The Condor* 68:5 (1966).

Belnap, Jayne. "The World at Your Feet: Desert Biological Soil Crusts." *Frontiers in Ecology and the Environment* 1 (2003): 181–89.

Benedict, Audrey D., and Joseph K. Gaydos. *The Salish Sea: Jewel of the Pacific Northwest*. Seattle: Sasquatch Books, 2015.

Bergland, Eric A. *Archaeological Test Excavations at the Berkeley Rockshelter Site (45-PI-303)*, Mt. Rainier National Park, Washington. SRD Report of Investigations No. 1, 1988.

Biek, David. *Flora of Mount Rainier National Park*. Corvallis: Oregon State University Press, 2000.

Bill, Will Jr. "Canoe Journey to Lummi 2019 is Now History." *Muckleshoot Messenger* 20, no. 5, August 8, 2019. www.muckleshoot.nsn.us/media/49275/muckleshoot messenger august 2019r.pdf.

Binford, Laurence C., Bruce G. Elliott, and Steven W. Singer. "Discovery of a Nest and the Downy Young of the Marbled Murrelet." *The Wilson Bulletin*, 87:3 (1975).

Blumenthal, Richard W. *Charles Wilkes and the Exploration of Inland Washington Waters: Journals from the Expedition of 1841.* Jefferson, NC: McFarland and Co., 2009.

Boeri, David J. "Studies on the Pollination of Sub-Alpine Plants at Mt. Rainier National Park." Master's thesis, University of Washington, 1976.

Bonar, Richard L. "Availability of Pileated Woodpecker Cavities and Use by Other Species." *The Journal of Wildlife Management,* 64:1 (2000), 52–59.

Bonney Lake Courier-Herald. "Lahar Warning System Receives Federal Funding." January 4, 2016. www.blscourierherald.com/news/364139251.html#.

Borror, Donald J. *Dictionary of Word Roots and Combining Forms.* Mountain View, CA: Mayfield Publishing Co., 1960.

Boyd, Robert T. *The Coming of the Spirit of Pestilence: Introduced Infectious Diseases and Population Decline Among Northwest Coast Indians, 1774–1874.* Seattle: University of Washington Press, 1999.

———, ed. *Indians, Fire and the Land in the Pacific Northwest.* Corvallis: Oregon State University Press, 1999.

Bradley, Russell W., Fred Cooke, Lynn W. Lougheed, and W. Sean Boyd. "Inferring Breeding Success through Radiotelemetry in the Marbled Murrelet." *Journal of Wildlife Management* 68 (2004): 318–31.

British Columbia Ministry of Forests, Lands and Natural Resource Operations. *Sustaining Communities: 2006–2011 Mountain Pine Beetle Action Plan, 2008 Progress Report.* www2.gov.bc.ca/assets/gov/farming-natural-resources-and-industry/forestry/forest-health/mountain-pine-beetle/mountain_pine_beetle_action_plan_progressreport_2008.pdf.

Brown, Rick, "Getting from 'No' to 'Yes': A Conservationist's Perspective." In Spies and Duncan, eds., *Old Growth in a New World.*

Bullock, Alison B. *The Flood of 2006: 2007 & 2008 Updates.* Unpublished, 2009. Ashford, WA: Mount Rainier National Park.

———. *Mount Rainier National Park Fact Sheet: Flood Recovery Status One Year Later.* Unpublished, 2007. Ashford, WA: Mount Rainier National Park.

———. "A Year after the Flood." Mount Rainier National Park news release. Unpublished, 2007. Ashford, WA: Mount Rainier National Park.

Burtchard, Greg, Ben Diaz, and Liz D'Arcy. "Archaeology and History in the Carbon River Valley Mount Rainier National Park." Draft, National Park Service, 2011.

Burtchard, Greg C., Jacqueline Y. Cheung, and Eric B. Gleason. *Archaeology and History in the Nisqually River Corridor.* Ashford, WA: Mount Rainier National Park, 2008.

Burtchard, Greg C., Stephen C. Hamilton, and Richard H. McClure Jr. *Environment, Prehistory and Archaeology of Mount Rainier National Park, Washington.* Seattle: National Park Service, 1998, repr. 2003.

Burton, Ken. "Creature Feature: Marbled Murrelet." *Econews,* 45:3 (2015). Arcata, CA: Northcoast Environmental Center.

Bush, Evan. "Mountain Goat Relocation Begins in Olympic National Park." *Seattle Times.* September 13, 2018. www.seattletimes.com/seattle-news/environment/mountain-goat-relocation-begins-in-olympic-national-park.

Cain, C. and S. Dolan. *Cultural Landscape Inventory, Carbon River Road, Mount Rainier National Park.* Unpublished draft, National Park Service, 2006. Ashford, WA: Mount Rainier National Park.

Cakir, Recep and Timothy J. Walsh. *Loss Estimation Pilot Project for Lahar Hazards from Mount Rainier, Washington.* Washington Division of Geology and Earth Resources, Information Circular 113, Olympia: Washington State Department of Natural Resources, 2012.

Cantwell, George G. "Notes on the Egg of the Marbled Murrelet." *The Auk,* 15:1 (1898), 49.

Carey, Andrew B. "Maintaining Biodiversity in Managed Forests." In Spies and Duncan, eds., *Old Growth in a New World.*

Carpenter, Cecilia Svinth. *The Nisqually—My People: The Traditional and Transitional History of the Nisqually People.* Tacoma, WA: Tahoma Research Service, 2002.

————. *Where the Waters Begin: The Traditional Nisqually Indian History of Mount Rainier*. Seattle, WA: Northwest Interpretive Association, 1994.

Carter, Harry R. and Spencer G. Sealy. "Who Solved the Mystery of the Marbled Murrelet?" *Northwestern Naturalist* 86 (2005): 2–11.

Catton, Theodore. *Wonderland: An Administrative History of Mount Rainier National Park*. Seattle, WA: National Park Service Cultural Resources Program, 1996.

————. *National Park, City Playground: Mount Rainier in the Twentieth Century*. Seattle: University of Washington Press, 2006.

Cederholm, C. Jeff, David H. Johnson, Robert E. Bilby, Lawrence G. Dominguez, Ann M. Garrett, William H. Graeber, Eva L. Greda, et al. "Pacific Salmon and Wildlife: Ecological Contexts, Relationships, and Implications for Management." In *Wildlife-Habitat Relationships in Oregon and Washington*. David H. Johnson and Thomas A. O'Neil. Corvallis: Oregon State University Press, 2001.

Center for Natural Lands Management. *Streaked Horned Lark Literature Review*. Unpublished report prepared for ENVIRON and Port of Portland, 2012.

Chadwick, Douglas H. *A Beast the Color of Winter*. San Francisco: Sierra Club Books, 1983.

————. *The Wolverine Way*. Ventura, CA: Patagonia Books, 2010.

Chanson, Hubert. "Sabo Check Dams—Mountain Protection Systems in Japan." In *International Journal of River Basin Management* 2:4 (2004), 301–7.

City of Tacoma, Washington. "The Foss Waterway Cleanup." 2015. www.cityoftacoma.org/ government/city_departments/environmentalservices/surface_water/restoration_and_ monitoring/thea_foss_waterway_cleanup.

Clement, Joel Pennington. "Structural Diversity and Epiphyte Distribution in Old-Growth Douglas Fir Tree Crowns." Master's thesis, The Evergreen State College, 1995.

Cloud Appreciation Society. "Your Turn To Do the Washing Up." August 3, 2009. cloudappreciationsociety.org/august-09.

Cohen, Fay G. *Treaties on Trial: The Continuing Controversy over Northwest Indian Fishing Rights*. Seattle: University of Washington Press, 1986.

Collier, Tracy Kay, Lindal L. Johnson, Mark S. Myers, Carla M. Stehr, Margaret M. Krahn, and John E. Stein. *Fish Injury in the Hylebos Waterway in Commencement Bay, Washington*. U.S. Dept. of Commerce, NOAA Tech. Memo. NMFS-NWFSC-36, 1998.

Coxson, D. S. and Nadkarni, N. M. "Ecological Roles of Epiphytes in Nutrient Cycles of Forest Ecosystems." In *Forest Canopies*, Margaret D. Lowman and Nalini M. Nadkarni, eds. San Diego: Academic Press, 1995.

Crandell, Dwight R. *The Geologic Story of Mount Rainier*. Geologic Survey Bulletin 1292. Washington, DC: U.S. Government Printing Office, 1969.

————. *Paradise Debris Flow at Mount Rainier, Washington*. U.S. Geological Survey Professional Paper, 475, B135-39, 1963.

————. *Postglacial Lahars from Mount Rainier Volcano, Washington*. Geological Survey Professional Paper 677. Washington, DC: U.S. Government Printing Office, 1971.

Crawford, Rex C. and Heidi Hall. "Changes in the South Puget Prairie Landscape." In *Ecology and Conservation of the South Puget Sound Prairie Landscape*. Patrick V. Dunn, and Kern Ewing, eds., 11–15. Seattle: The Nature Conservancy of Washington, 1997.

Crawford, Rodney L. and John S. Edwards. "Ballooning Spiders as a Component of Arthropod Fallout on Snowfields of Mount Rainier, Washington." *USA Arctic and Alpine Research*, 18:4 (1986), 429–37.

Czuba, Jonathan A., Christopher S. Magirl, Christiana R. Czuba, Christopher A. Curran, Kenneth H. Johnson, Theresa D. Olsen, Halley K. Kimball, and Casey C. Gish. *Geomorphic Analysis of the River Response to Sedimentation Downstream of Mount Rainier, Washington*. U.S. Geological Survey Open-File Report 2012-1242.

Czuba, Jonathan A., Christopher S. Magirl, Christiana R. Czuba, Eric E. Grossman, Christopher A. Curran, Andrew S. Gendaszek, and Richard S. Dinicola. *Sediment Load from Major Rivers into Puget Sound and Its Adjacent Waters.* Fact Sheet 2011-3083. U.S. Department of the Interior, U.S. Geological Survey. Washington, DC: U.S. Government Printing Office, 2011.

Dale, Virginia H., Linda A. Joyce, Steve McNulty, Ronald P. Neilson, Matthew P. Ayres, Michael D. Flannigan, Paul J. Hanson, Lloyd C. Irland, Ariel E. Lugo, Chris J. Peterson, Daniel Simberloff, Frederick J. Swanson, Brian J. Stocks, B. Michael Wotton. "Climate Change and Forest Disturbances." *BioScience* 51:9 (2001): 723–34.

Dalquest, Walter Woelber. *Mammals of Washington.* University of Kansas Museum of Natural History Publication. Lawrence: University of Kansas, 1948.

Daugherty, Richard D. *An Archaeological Survey of Mount Rainier National Park.* Manuscript on file at Mount Rainier National Park, 1963.

Day, John A. *The Book of Clouds.* New York: Silver Lining Books, 2003.

De Laguna, Frederica. *Under Mount Saint Elias: The History and Culture of the Yakutat Tlingit.* Washington, DC: Smithsonian Institution Press, 1972.

Desimone, S. M. *Draft Periodic Status Review for the Marbled Murrelet in Washington.* Olympia, WA: Washington Department of Fish and Wildlife, 2016.

Dhundale, J. *Marbled Murrelet 2009 Progress Report Mount Rainier National Park.* Unpublished, 2009.

Dietrich, William. *The Final Forest: The Battle for the Last Great Trees of the Pacific Northwest.* New York: Penguin Books, 1993.

Diller, Lowell V. "To Shoot Or Not To Shoot: The Ethical Dilemma of Killing One Raptor To Save Another." *The Wildlife Professional,* Winter 2013. www.orww.org/Elliott_Forest/References/Academic/Wildlife_Society/Diller_20130100.pdf.

Donkor, Noble T. "Impact of Beaver (*Castor canadensis Kuhl*) Foraging on Species Composition of Boreal Forests." In *Plant Disturbance Ecology: The Process and The Response,* Edward Arnold Johnson and Kiyoko Miyanishi, eds. Amsterdam: Elsevier Academic Press, 2007.

Douglas, David. *Journal Kept by David Douglas during His Travels in North America 1823–1827.* New York: Antiquarian Press, 1959.

Dragovich, Joe D., Patrick T. Pringle, and Timothy J. Walsh. "Extent and Geometry of the Mid-Holocene Osceola Mudflow in the Puget Lowland—Implications for Holocene Sedimentation and Paleogeography." *Washington Geology* 22:3 (1994), 3–26.

Driedger, Carolyn L. *A Visitor's Guide to Mount Rainier Glaciers.* Longmire, WA: Pacific Northwest Parks and Forest Association, 1986.

Driedger, Carolyn L. and Andrew G. Fountain. "Glacier Outburst Floods at Mount Rainier, Washington State, USA." In B. Wold, ed., *Annals of Glaciology: Proceedings of the Symposium on Snow and Glacier Research Relating To Human Living Conditions,* vol. 13 (1989): 51–55.

Driedger, Carolyn L. and K. M. Scott. *Mount Rainier—Learning To Live With Volcanic Risk.* Fact Sheet 034-02. United States Geological Survey, 2002.

Driedger, Carolyn L. and Paul M. Kennard. *Ice Volumes on Cascade Volcanoes—Mount Rainier, Mount Hood, Three Sisters, and Mount Shasta.* U.S. Geological Survey Professional Paper 1365, 1986.

Driedger, Carolyn L. and William E. Scott. *Mount Rainier—Living Safely With a Volcano in Your Backyard.* Fact Sheet 2008-3062. United States Geological Survey, 2008.

Duffield, W. James. "Pollination Ecology of Castilleja in Mount Rainier National Park." *Ohio Journal of Science,* 72:2 (1972), 110–14.

Dugger, Katie M., Eric D. Forsman, Alan B. Franklin, Raymond J. Davis, et al. "The Effects of Habitat, Climate, and Barred Owls on Long-Term Demography of Northern Spotted Owls." *The Condor: Ornithological Applications* 118 (2016): 57–116.

Duncan, Sally. "Coming Home to Roost: The Pileated Woodpecker as Ecosystem Engineer." In *Science Findings,* Issue 57. U.S. Dept. of Agriculture. Portland, OR: Pacific Northwest Research Station, 2003.

Duncan, Sally L., Denise Lach, and Thomas A. Spies. "Old Growth in a New World: A Synthesis." In Spies and Duncan, eds., *Old Growth in a New World.*

Dunwiddie, Peter W. and Bakker, Jonathan D. "The Future of Restoration and Management of Prairie-Oak Ecosystems in the Pacific Northwest." *Northwest Science* 85:2 (2009), 83–92.

Dzurisin, Daniel, Carolyn L. Driedger, and Lisa M. Faust, *Mount St. Helens, 1980 to Now—What's Going On?* U.S. Geological Survey Fact Sheet 2013–3014, 2013.

Earth Economics. *The Puyallup River Watershed: An Ecological Economic Characterization.* 2011. issuu. com/earth_economics/docs/puyallup_river_watershed.

Edwards, Ola Margery. "The Alpine Vegetation of Mount Rainier National Park: Structure, Development, and Constraints." PhD thesis, University of Washington, 1980.

Elliott, Carl. "So Close to a Million Plants We Can Almost Taste It." Sustainability in Prisons Project. February 24, 2014. sustainabilityinprisons.org/blog/2014/02/25/so-close-to-a-million-plants-we-can-almost-taste-it.

Ellison, Aaron M., Michael S. Bank, Barton D. Clinton, Elizabeth A. Colburn, Katherine Elliott, Chelcy R. Ford, David R. Foster, Brian D. Kloeppel, Jennifer D. Knoepp, Gary M. Lovett, Jacqueline Mohan, David A. Orwig, Nicholas L. Rodenhouse, William V. Sobczak, Kristina A. Stinson, Jeffrey K. Stone, Christopher M. Swan, Jill Thompson, Betsy Von Holle, and Jackson R. Webster. "Loss of Foundation Species: Consequences for the Structure and Dynamics of Forested Ecosystems." *Frontiers in Ecology* 3:9 (2005), 479–86.

Elphick, Chris, John B. Dunning Jr., and David Allen Sibley. *The Sibley Guide to Bird Life and Behavior.* New York: Alfred A. Knopf, 2001.

Ettl, Gregory J. and Nicholas Cottone. "Whitebark Pine (*Pinus albicaulis*) in Mt. Rainier National Park, Washington, USA: Response to Blister Rust Infection." In *Species Conservation and Management: Case Studies*, 36–47. H. Resit Akçayaya, Mark A. Burgman, and Oskar Kindvall, eds. Oxford: Oxford University Press, 2004.

Evans Mack, Diane, William P. Ritchie, S. Kim Nelson, Elena Kuo-Harrison, Peter Harrison, and Thomas E. Hamer. *Methods for Surveying Marbled Murrelets in Forests: A Revised Protocol for Land Management and Research.* Pacific Seabird Group Technical Publication Number 2, 2003. pacificseabirdgroup.org/psg-publications/technical-publications.

Falxa, Gary A. and Martin G. Raphael, *Northwest Forest Plan—The First Twenty Years (1994–2013): Status and Trend of Marbled Murrelet Populations and Nesting Habitat.* General Technical Report PNW-GTR-933. Portland, OR: U.S. Department of Agriculture, Forest Service, Pacific Northwest Research Station, 2016.

Festa-Bianchet, Marco and Steeve D. Côtè. *Mountain Goats: Ecology, Behavior, and Conservation of an Alpine Ungulate.* Washington, DC: Island Press, 2008.

Finn, Carol A., Thomas W. Sisson, and Marla Deszcz-Pan. "Aerogeophysical Measurements of Collapse-Prone Hydrothermally Altered Zones at Mount Rainier Volcano." *Nature* 409 (2001), 600–603.

Fiore, David C., Denise D. McKee, and Mark A. Janiga. "Red Snow: Is it Safe to Eat? A Pilot Study." *Wilderness and Environmental Medicine* 8 (1997), 94–95.

Fiske, Richard S., Clifford A. Hopson, and Aaron C. Waters. *Geology of Mount Rainier National Park Washington.* Geologic Survey Professional Paper 444. Washington, DC: U.S. Government Printing Office, 1963.

Floodplains by Design. "Floodplains by Design: Reducing Risk, Restoring Rivers." www.floodplains bydesign.org.

Fogel, Elise. Nisqually Delta Association. History Link.org. October 3, 2011. www.historylink.org/File/9940.

Forsman, Eric D., Robert G. Anthony, Katie M. Dugger, Elizabeth M. Glenn, Allen B. Franklin, Gary C. White, Carl J. Schwarz, Kenneth B. Burnham, David R. Anderson, and James D. Nichols. "Population Demography of Northern Spotted Owls." *Studies in Avian Biology* 40:40 (2011).

Franklin, Jerry F. "Conserving Old-Growth Forests and Attributes: Reservation, Restoration, and Resilience." In Spies and Duncan, eds., *Old Growth in a New World.*

Franklin, Jerry F. and C. T. Dyrness. *Natural Vegetation of Oregon and Washington.* Portland, OR: Pacific Northwest Forest and Range Experiment Station, Forest Service, U.S. Department of Agriculture, 1973.

Franklin, Jerry F., Kermit Cromack Jr., William Denison, Arthur McKee, Chris Maser, James Sedell, Fred Swanson, and Glen Juday. *Ecological Characteristics of Old-Growth Douglas-Fir Forests.* USDA, Forest Service. Pacific Northwest Forest and Range Experimental Station. General Technical Report PNW-118, 1981.

Fraser, Alistair B. "Laminated Cap and Wave Clouds." *Weatherwise,* 31:2 (1978).

Gallacci, Caroline Denyer and Karabaich, Ron. *The City of Destiny and the South Sound: An Illustrated History of Tacoma and Pierce County.* Carlsbad, CA: Heritage Media, 2001.

Geils, Brian W. and Detlev R. Vogler. *A Natural History of Cronartium ribicola.* USDA Forest Service Proceedings RMRS-P-63, 2011.

Gibbons, Bob. *Wildflower Wonders: The 50 Best Wildflower Sites in the World.* Princeton, NJ: Princeton University Press, 2011.

Gibbs, George. *Tribes of Western Washington and Northwestern Oregon.* Department of the Interior: U.S. Geographical and Geological Survey of the Rocky Mountain Region, 1877.

Groc, Isabelle. "Shooting Owls to Save Other Owls." *National Geographic Online.* July 17, 2014. news.nationalgeographic.com/news/2014/07/140717-spotted-owls-barred-shooting-logging-endangered-species-science.

Guiguet, C. J. "Enigma of the Pacific." *Audubon* 58 (1956): 164-167, 174.

Gunther, Erna. "An Analysis of the First Salmon Ceremony." *American Anthropologist* 28 (1926): 605–17.

———. "A Further Analysis of the First Salmon Ceremony." *University of Washington Publications in Anthropology* 2, no. 5 (1928).

———. "Ethnobotany of Western Washington: The Knowledge and Use of Indigenous Plants by Native Americans." Rev. ed. Seattle: University of Washington Press, 1973.

Haeberlin, Hermann and Erna Gunther. *The Indians of Puget Sound.* Seattle: University of Washington Press, 1930.

Haines, Aubrey L. *Mountain Fever: Historic Conquests of Rainier.* Seattle: University of Washington Press, 1999.

Halliday, William R. and Charles H. Anderson Jr. *The Paradise Ice Caves, Mount Rainier National Park Washington.* N.p., 1972.

Hamel, Nathalie, Jerry Joyce, Mindy Fohn, Andy James, Jason Toft, Alicia Lawver, Scott Redman, and Marie Naughton, eds. *2015 State of the Sound: Report on the Puget Sound Vital Signs.* November 2015. www.psp.wa.gov/sos.

Hamman, Sarah T., Peter W. Dunwiddie, Jason L. Nuckols, and Mason McKinley. "Fire as a Restoration Tool in Pacific Northwest Prairies and Oak Woodlands: Challenges, Successes and Future Directions." *Northwest Science* 85:2 (2011), 317–28.

Haristoy, Camila Tejo, Darlene Zabowski, and Nalini Nadkarni. "Canopy Soils of Sitka Spruce and Bigleaf Maple in the Queets River Watershed, Washington." *Soil Science Society of America Journal* 78, no. S1 (2014).

Harke, Vince. "Mount Rainier National Park Distribution of Marbled Murrelet (MAMU) Habitat and Survey Stations." Unpublished report, 2009. Lacey, WA: U.S. Fish and Wildlife Service.

Harmon, Alexandra. *Indians in the Making: Ethnic Relations and Indian Identities around Puget Sound.* Berkeley: University of California Press, 1998.

Harris, Stephen L. *Fire and Ice: The Cascade Volcanoes.* Seattle: Mountaineers, 1976.

———. *Fire Mountains of the West: The Cascade and Mono Lake Volcanoes.* Missoula: Mountain Press, 1988.

Hart Crowser. "Sitcum/Blair/Milwaukee Waterway Nearshore Fill." 2015.

Hartman, John Peter. *Creation of Mount Rainier National Park*. Address delivered at the 37th Annual Convention of Washington Good Roads Association, 1935.

Hatfield, R., S. Jepsen, R. Thorp, L. Richardson, S. Colla, and S. Foltz Jordan. 2015. "Western Bumblebee: *Bombus occidentalis.*" *The IUCN Red List of Threatened Species 2015*. dx.doi.org/10.2305/IUCN.UK.2015-2.RLTS.T44937492A46440201.en.

Heffernan, Trova. *Where the Salmon Run: The Life and Legacy of Billy Frank Jr.* Seattle: University of Washington Press, 2012.

Helfield, James M. and Robert J. Naiman. "Effects of Salmon-Derived Nitrogen on Riparian Forest Growth and Implications for Stream Productivity." *Ecology* 82, no. 9 (September 2001), 2403–9.

Henderson, Jan Alan. "Composition, Distribution and Succession of Subalpine Meadows in Mount Rainier National Park." PhD thesis, Oregon State University, 1973.

Heter, Elmo W. "Transplanting Beavers by Airplane and Parachute." *Journal of Wildlife Management* 14:2, 1950.

Hilbert, Vi, ed. *Haboo: Native American Stories from Puget Sound.* Seattle: University of Washington Press, 1985.

Hilbert, Vi, Jay Miller, and Zalmai Zahir, eds. *Puget Sound Geography: Original Manuscript from T. T. Waterman.* Federal Way, WA: Zahir Consulting Services, 2001.

Hobbs, Jared. "Habitat Action Plan for Northern Spotted Owl Pursuant to the Species at Risk Act." (Letter from Ecojustice to Catherine McKenna, Canadian Minister of Environment, May 8, 2019). www.wildernesscommittee.org/news/environmental-groups-demand-feds-act-save-critically-endangered-spotted-owl.

Hoblitt, R. P., J. S. Walder, C. L. Driedger, K. M. Scott, P. T. Pringle, and J. W. Vallance. *Volcano Hazards from Mount Rainier, Washington.* Open-File Report 98-428. U.S. Department of the Interior, U.S. Geological Survey, revised 1988.

Hoham, Ronald W. and Brian Duval. "Microbial Ecology of Snow and Freshwater Ice with Emphasis on Snow Algae." In *Snow Ecology: An Interdisciplinary Examination of Snow-Covered Ecosystems.* H. Gerald Jones, J. W. Pomeroy, D. A. Walker, and R. W. Hoham, eds. Cambridge: Cambridge University Press, 2001.

Howard, Calvin D. "The Atlatl: Function and Performance." *American Antiquity* 39:1 (1974), 102–4.

Hunn, Eugene S. and David H. French. "Lomatium: A Key Resource for Columbia Plateau Native Subsistence." *Northwest Science* 55:2 (1981), 87–94.

Hunn, Eugene S. and James Selam. *Nch'i-Wána "The Big River": Mid-Columbia Indians and Their Land.* Seattle: University of Washington Press, 1990.

Intergovernmental Panel on Climate Change. *Climate Change 2007: Impacts, Adaptation and Vulnerability. Contribution of Working Group II to the Fourth Assessment Report of the Intergovernmental Panel on Climate Change.* Cambridge: Cambridge University Press, 2007.

Jenkins, Kurt, Katherine Beirne, Patricia Happe, Roger Hoffman, Cliff Rice, and Jim Schaberl. *Seasonal Distribution and Aerial Surveys of Mountain Goats in Mount Rainier, North Cascades, and Olympic National Parks, Washington.* U.S. Geological Survey Open-File Report 2011-1107 (2011).

John, David A., Thomas W. Sisson, George N. Breit, Robert O. Rye, and James W. Vallance. "Characteristics, Extent and Origin of Hydrothermal Alteration at Mount Rainier Volcano, Cascades Arc, USA: Implications for Debris-Flow Hazards and Mineral Deposits." *Journal of Volcanology and Geothermal Research*, 175:3 (2008).

Johnson, Elden. "Book Review: The Atlatl in North America." *American Antiquity* 22, no. 1 (1956): 86.

Johnson, K. Norman and Frederick J. Swanson. "Historical Context of Old-Growth Forests in the Pacific Northwest: Policy, Practices and Competing Worldviews." In Spies and Duncan, eds., *Old Growth in a New World.*

Jolley, L. W. and W. R. English. *What Is Fecal Coliform? Why Is It Important?* Clemson University Cooperative Extension Forestry and Natural Resources (2015). www.clemson.edu/extension/natural_resources/water/publications/fecal_coliform.html.

Jones, Lawrence L. C., William P. Leonard, and Deanna H. Olson, eds. *Amphibians of the Pacific Northwest*. Seattle: Seattle Audubon Society, 2005.

Jost, Robby, Jeni Chan, and Paul Kennard. "The Westside Road: Physical Flood Protection in Mount Rainier National Park, Washington." Presentation at Geological Society of America Annual Meeting, Seattle, WA, 2017. Also see Beason et al. *Rivers Gone Wild*.

Judge, Millie M. "2011 Implementation Status Assessment Final Report: A Qualitative Assessment of Implementation of the Puget Sound Chinook Recovery Plan." Prepared for the National Marine Fisheries Services. Task Order 2002. Seattle: National Marine Fisheries Services. archive.fisheries.noaa.gov/wcr/protected_species/salmon_steelhead/recovery_planning_and_implementation/puget_sound/puget_sound_chinook_recovery_plan.html.

Kautz, August V. "Ascent of Mount Rainier." *The Overland Monthly* 14, no. 5 (May 1875).

Kays, Roland W. and Don E. Wilson. *Mammals of North America*. Princeton, NJ: Princeton University Press, 2002.

Keane, Robert E. and Stephen F. Arno, "Restoration Concepts and Techniques." In Tomback et al., *Whitebark Pine Communities*.

Kerr, Andy. "Starting the Fight and Finishing the Job." In Spies and Duncan, eds., *Old Growth in a New World* (2009).

Kerwin, John. *Salmon Habitat Limiting Factors Report for the Puyallup River Basin Water Resource Inventory Area 10*. Olympia, WA: Washington Conservation Commission, 1999.

Kimmerer, Robin Wall. *Gathering Moss: A Natural and Cultural History of Mosses*. Corvallis: Oregon State University Press (2003).

King County, Washington. *2013 Flood Hazard Management Plan Update: King County, Washington*. King County Department of Natural Resources and Parks, Water and Land Resources Division. Seattle, Washington, 2013.

———. *Lower White River Countyline Levee Setback Project*. 2016. www.kingcounty.gov/depts/dnrp/wlr/sections-programs/river-floodplain-section/capital-projects/lower-white-river-countyline-a-street.aspx.

Kirk, Ruth. *Sunrise to Paradise: The Story of Mount Rainier National Park*. Seattle: University of Washington Press, 1999.

Kirk, Ruth and Richard D. Daugherty. *Archaeology in Washington*. Seattle: University of Washington Press, 2007.

Kopperl, Robert, Charles Hodges, Christian Miss, Johonna Shea, and Alecia Spooner. *Archaeology of King County, Washington: A Context Statement for Native American Archaeological Resources*. Seattle: SWCA Environmental Consultants. Prepared for the King County Historic Preservation Program, Seattle WA, 2016.

Kronland, Bill. *Mazama Pocket Gopher Response to Prescribed Fire on Joint Base Lewis-McChord*. Presentation at Cascadia Prairie Oak Partnership Conference. Eugene, Oregon, 2018.

Kruckeberg, Arthur R. *The Natural History of Puget Sound Country*. Seattle: University of Washington Press, 1995.

Kyle, C. J., T. J. Karels, C. S. Davis, S. Mebs, B. Clark, C. Strobeck, et al. "Social Structure and Facultative Mating Systems of Hoary Marmots (*Marmota caligata*)." *Molecular Ecology* 16:6 (2007), 1245–55.

Lakatos, Michael and Alexandra Fischer-Pardow. "Nonvascular Epiphytes: Functions and Risks at the Tree Canopy." In *Treetops at Risk: Challenges of Global Canopy Ecology and Conservation*. Margaret Lowman, Soubadra Devy, and T. Ganesh, eds. New York: Springer, 2013.

Lang, George. *Making Wawa: The Genesis of Chinook Jargon*. Vancouver: University of British Columbia Press, 2008.

Lanner, Ronald M. "Adaptations of Whitebark Pine for Seed Dispersal by Clark's Nutcracker. *Canadian Journal of Forest Research* 12 (1982): 391–402.

Le, Phuong. "U.S. Prison Inmates Raise Rare Frogs, Butterflies." *USA Today*. September 22, 2012. usatoday30.usatoday.com/news/nation/story/2012/09/22/us-prison-inmates-raise-rare-frogs-butterflies/57827452/1.

————. "NOAA: Fish Ladder Required at Mud Mountain Dam." *Seattle Times*. October 7, 2014. www.seattletimes.com/seattle-news/noaa-fish-ladder-required-at-mud-mountain-dam.

Lee, Jessica. "Mount Rainier to Get New Digital-Warning System for Massive Mudflows." *Seattle Times*. January 2, 2017. www.seattletimes.com/seattle-news/environment/mount-rainier-to-get-new-digital-warning-system-for-massive-mudflows.

Lee, Robert G. "Sacred Trees." In Spies and Duncan, eds., *Old Growth in a New World*.

Legg, Nicholas T., Andrew J. Meigs, Gordon E. Grant, and Paul Kennard. "Debris Flow Initiation in Proglacial Gullies on Mount Rainier, Washington." *Geomorphology* 226 (2014): 249–60.

Leopold, Aldo. *A Sand County Almanac and Sketches Here and There*. London: Oxford University Press, 1949.

Leopold, Estella B. and Robert Thomas Boyd. "An Ecological History of Old Prairie Areas in Southwestern Washington." In *Indians, Fire and the Land in the Pacific Northwest*. Robert Thomas Boyd, ed. Corvallis: Oregon State University Press, 1999.

Lescinsky, D. T. and T. W. Sisson. "Ridge-Forming, Ice-Bounded Lava Flows at Mount Rainier, Washington." *Geology* 26:4 (1999), 351–54.

Lichatowich, Jim. *Salmon Without Rivers: A History of the Pacific Salmon Crisis*. Washington, DC: Island Press, 1999.

————. *Salmon, People and Place: A Biologist's Search for Salmon Recovery*. Corvallis: Oregon State University Press, 2013.

Lillie, Robert J. *Parks and Plates: The Geology of Our National Parks, Monuments and Seashores*. New York: W. W. Norton, 2005.

Logan, Jesse A., Jacques Régnière, and James A. Powell. "Assessing the Impacts of Global Warming on Forest Pest Dynamics." *Frontiers in Ecology and the Environment*, 1: 130–37.

Lowery, David P. *Western Red Cedar: An American Wood*. U.S. Department of Agriculture. FS-261. U.S. Government Printing Office, 1984.

Loyd, Nic and Linda Weiford, "Lenticular Clouds—The Truth Is Out There." WSU News, September 14, 2016. news.wsu.edu/2016/09/14/lenticular-clouds-truth.

Lubinski, Patrick M. and Greg C. Burtchard. "Archaeology of Fryingpan Rockshelter (45PI43) in Mount Rainier National Park." *Archaeology in Washington*, 11 (2005).

Luyssaert, Sebastiaan, Beverly E. Law, Ernst-Detlef Schulze, Annette Börner, Alexander Knohl, Dominik Hessenmöller, Philippe Ciais, and John Grace. "Old-Growth Forests as Global Carbon Sinks." *Nature* 455, no. 7210 (2008): 213–15.

Macior, Lazarus Walter. "The Pollination Ecology of Pedicularis on Mount Rainier." *American Journal of Botany* 60:9 (1988).

————. "Floral Resource Sharing by Bumblebees and Hummingbirds in Pedicularis (Scrophulariaceae)." *Bulletin of the Torrey Botanical Club*, 113:2 (1986), 101–9.

Mack, Cheryl A. "A Burning Issue: American Indian Fire Use on the Mt. Rainier Forest Reserve." *Fire Management Today* 62:2 (2003), 20–24.

Mack, Cheryl A., and Richard H. McClure Jr. "Vaccinium Processing in the Washington Cascades." *Journal of Ethnobiology*, 22:1 (2002), 35–60.

Mahalovich, M., and L. Stritch. "Whitebark Pine: *Pinus albicaulis*." *The IUCN Red List of Threatened Species 2013*. dx.doi.org/10.2305/IUCN.UK.2013-1.RLTS.T39049A2885918.en.

Mapes, Lynda V. "After 153 Years, Treaty Tree Lost to Winter Storm." *Seattle Times*. February 12, 2007. www.seattletimes.com/seattle-news/after-153-years-treaty-tree-lost-to-winter-storm.

————. "Quake Monitors on Mount Rainier Ready and Waiting." *Seattle Times*. September 16, 2012. www.seattletimes.com/seattle-news/quake-monitors-on-mount-rainier-ready-and-waiting.

Marks, E. L., R. C. Ladley, B. E. Smith, A. G. Berger, J. A. Paul, T. G. Sebastian, and K. Williamson. *2014–2015 Annual Salmon, Steelhead, and Bull Trout Report: Puyallup/White River Watershed—Water Resource Inventory Area 10*. Puyallup, WA: Puyallup Tribal Fisheries, 2015.

Martin, John. "Different Feeding Strategies of Two Sympatric Hummingbird Species." *The Condor* 90 (1988): 233–36.

Maser, Chris. *Mammals of the Pacific Northwest: From the Coast to the High Cascades.* Corvallis: Oregon State University Press, 1998.

Maser, Chris and James M. Trappe. *Seen and Unseen World of the Fallen Tree.* USDA Forest Service, Pacific Northwest Forest and Range Experiment Station, General Technical Report PNW-164, 1984.

Mass, Cliff. *The Weather of the Pacific Northwest.* Seattle: University of Washington Press, 2008.

Mathews, Daniel. *Cascade-Olympic Natural History: A Trailside Reference,* 2nd ed. Portland, OR: Raven Editions, 1999.

Matrazzo, Donna. *Wildlife in Managed Forests: Northern Spotted Owl.* Portland: Oregon Forest Resources Institute, 2007.

Mattson, David J., Katherine C. Kendall, and Daniel P. Reinhart. "Whitebark Pine, Grizzly Bears and Red Squirrels." In Tomback et al., *Whitebark Pine Communities.*

Mauger, Guillaume S., Joseph H. Casola, Harriet A. Morgan, Ronda L. Strauch, Brittany Jones, Beth Curry, Tania M. Busch Isaksen, Lara Whitely Binder, Meade B. Krosby, and Amy K. Snover. (2015). *State of Knowledge: Climate Change in Puget Sound.* Report prepared for the Puget Sound Partnership and the National Oceanic and Atmospheric Administration. Climate Impacts Group, University of Washington. Seattle, 2015. doi:10.7915/CIG93777D.

Mayor, Jeffrey P. "Unearthing Secrets of the Ancient Cascades." *Tacoma News Tribune.* November 15, 2009. tdn.com/news/scientists-unearthing-secrets-of-the-ancient-cascades/article_6a26b766-b549-5328-a2a5-85a3eab7d9b3.html.

———. "A Mountain in Transit." *Olympian* (Olympia, WA). December 5, 2010.

———. "Rocks on the Move." *Olympian* (Olympia, WA). December 5, 2010.

———. "Carbon River Road Transition to Trail Begins." *Tacoma News Tribune.* July 24, 2011.

———. (2013). "Mount Rainier Volunteer Program Sees Drop in Hours Worked in 2013." *Olympian* (Olympia, WA).

McCune, Bruce, Linda Geiser, Sylvia Duran Sharnoff, Stephen Sharnoff, and Alexander G. Mikulin. *Macrolichens of the Pacific Northwest.* Corvallis: Oregon State University Press, 2009.

McCune, B., K. A. Amsberry, and M. Widmer. "Vertical Profile of Epiphytes in a Pacific Northwest Old-Growth Forest." *Northwest Science,* 71:2 (1997), 145–52.

McCune, Bruce, Roger Rosentreter, Jeanne M. Ponzetti, and David C. Shaw, "Epiphyte Habitats in an Old Conifer Forest in Western Washington, USA." *The Bryologist,* 103:3 (2000), 417–27.

McCutcheon, Patrick T. "Enhancing Visitor Experience and Appreciation of Mount Rainier Archaeology through Excavations at 45PI408: Phase I, Research Design." Unpublished. Ellensburg: Central Washington University, 2011.

McDonald, Geral I. and Raymond J. Hoff. "Blister Rust: An Introduced Plague." In Tomback et al., *Whitebark Pine Communities.*

McNulty, Tim. *Olympic National Park: A Natural History.* Seattle: University of Washington Press, 2009.

McWhorter, Lucullus V. "Chief Sluskin's True Narrative." *Washington Historical Quarterly* 8, no. 2 (1917).

Meany, Edmond S. *Mount Rainier: A Record of Exploration.* New York: Macmillan Company, 1916.

———. *Origin of Washington Geographic Names.* Seattle: University of Washington Press, 1923.

Meeker, Ezra. *The Tragedy of Leschi.* Seattle: Historical Society of Seattle and King County, 1980.

Miles, John C. "Wilderness in National Parks: Playground or Preserve." Seattle: University of Washington Press, 2009.

Miller, Sherri L, Martin G. Raphael, Gary A. Falxa, Craig Strong, Jim Baldwin, Thomas Bloxton, Beth M. Galleher, Monique Lance, Deanna Lynch, Scott F. Pearson, C. John Ralph, and Richard D. Young. "Recent Population Decline of the Marbled Murrelet in the Pacific Northwest." *The Condor* 114:4 (2012), 771–81.

Minore, Don. *Western Redcedar—A Literature Review.* USDA Forest Service, Pacific Northwest Forest and Range Experimental Station General Technical report PNW-150, 1983.

Moir, William H. *Forests of Mount Rainier National Park: A Natural History*. Seattle: The Pacific Northwest National Parks and Forests Association, 1989.

Molenaar, Dee. *The Challenge of Rainier*. Seattle: Mountaineers, 1971.

Møller, Anders Pape, Wolfgang Fiedler, and Peter Berthold. *Effects of Climate Change on Birds*. Oxford: Oxford University Press, 2010.

Moore, Kathleen Dean. "In the Shadow of the Cedars: Spiritual Values of Old-Growth Forests." In Spies and Duncan, eds., *Old Growth in a New World*.

Morgan, Murray. *Puget's Sound: A Narrative of Early Tacoma and the Southern Sound*. Seattle: University of Washington Press, 2003.

Moskowitz, David. "Wildlife of the Pacific Northwest: Tracking and Identifying Mammals, Birds, Reptiles, Amphibians, and Invertebrates." Portland, OR: Timber Press, 2010.

Mosser, J. L., A. G. Mosser, and T. D. Brock. "Photosynthesis in the Snow: The Alga Chlamydomonas Nivalis (*Chlorophyceae*)." *Journal of Phycology* 13 (1977): 22–27.

Moulton, Gary E., ed. *The Journals of the Lewis and Clark Expedition*, vol. 6. Lincoln: University of Nebraska Press, 1983.

Mudd Ruth, Maria. *Rare Bird: Pursuing the Mystery of the Marbled Murrelet*. Emmaus, PA: Rodale, 2005.

Müller-Schwarze, Dietland, and Lixing Sun. *The Beaver: Natural History of a Wetlands Engineer*. Ithaca, NY: Cornell University Press, 2003.

Myers, Ellen. *Final Report Mount Rainier National Park: Marbled Murrelet 2003 Progress Report*. Unpublished, 2004.

Nadkarni, Nalini M. *Between Earth and Sky: Our Intimate Connection to Trees*. Berkeley: University of California Press, 2008.

———. "Biomass and Mineral Capital of Epiphytes in an *Acer macrophyllum* Community of Temperate Moist Coniferous Forest, Olympic Peninsula, Washington State." *Canadian Journal of Botany* 62, no. 11 (1984): 2223–28.

———. "Canopy Roots: Convergent Evolution in Rainforest Nutrient Cycles." *Science* 214:4524 (1981).

———. "Roots That Go Out on a Limb." *Natural History*, 94:5 (1985), 42–49.

Nadkarni, Nalini M., Mark C. Merwin, and Jurgen Nieder. "Forest Canopies, Plant Diversity." In *Encyclopedia of Biodiversity*, vol. 3. San Diego: Academic Press, 2001.

National Marine Fisheries Service. *The Endangered Species Act —Protecting Marine Resources*. National Oceanic and Atmospheric Administration, 2015.

National Oceanic and Atmospheric Administration (NOAA) Fisheries. *Chinook Salmon (Oncorhynchos tshawytscha)*. www.fisheries.noaa.gov/species/chinook-salmon.

National Park Service. *Olympic National Park Mountain Goat Action Plan*. Unpublished, 2011.

———. *Mountain Goat Capture and Translocation*. 2018. www.nps.gov/olym/planyourvisit/mountain-goat-capture-and-translocation.htm.

National Wildlife Federation. *Wildife Guide: Chinook Salmon*. www.nwf.org/sitecore/content/Home/Educational-Resources/Wildlife-Guide/Fish/Chinook-Salmon.

Neiman, Paul J., F. Martin Ralph, Gary A. Wick, Ying-Hwa Kuo, Tae-Kwon Wee, Zaizhong Ma, George H. Taylor, and Michael D. Dettinger. "Diagnosis of an Intense Atmospheric River in the Pacific Northwest: Storm Summary and Offshore Vertical Structure Observed with Cosmic Satellite Retrievals." *Monthly Weather Review*, 136:11 (2008), 4398–4420.

Neiman, Paul J., Lawrence J. Schick, F. Martin Ralph, Mimi Hughes, and Gary A. Wick. "Flooding in Western Washington: The Connection to Atmospheric Rivers." *Journal of Hydrometeorology*, 12:6 (2011), 1337–58.

Nisbet, Jack. *The Collector: David Douglas and the Natural History of the Northwest*. Seattle: Sasquatch Books, 2009.

Nisqually Land Trust. "Mount Rainier Gateway." 2016. nisquallylandtrust.org/our-lands-and-projects/protected-areas/mount-rainier-gateway-initiative.

————. "Nisqually Community Forest." 2016. nisquallylandtrust.org/our-lands-and-projects/nisqually-community-forest.

Nisqually National Wildlife Refuge. "History." www.fws.gov/refuges/profiles/History.cfm?ID=13529.

Nisqually River Council. Nisqually Watershed Stewardship Plan. 2011. www.slideshare.net/Nisqually/nisqually-watershed-stewardship-plan.

Noon, Barry R. "Old-Growth Forest as Wildlife Habitat." In Spies and Duncan, eds., *Old Growth in a New World*.

Northwest Indian Fisheries Commission. *State of Our Watersheds Report*. Olympia, WA: Northwest Indian Fisheries Commission, 2012.

Northwest Treaty Tribes. "Logjams on the Mashel River Help Fish, Protect Property." *NW Treaty Tribe News*. September 18, 2009. nwtreatytribes.org/logjams-on-the-mashel-river-help-fish-protect-property.

————. "Making Way for Chinook on Ohop Creek." *Northwest Treaty Tribe News*, September 11, 2009. nwtreatytribes.org/making-way-for-chinook-on-ohop-creek.

————. "New Mashel River Logjams Protect Salmon and Property." *Northwest Treaty Tribe News*, September 7, 2010. nwtreatytribes.org/new-mashel-river-logjams-protect-salmon-and-property.

————. "Nisqually Tribe Tracking Salmon through Ohop Creek." *Northwest Treaty Tribe News*, October 18, 2012. nwtreatytribes.org/nisqually-tribe-tracking-salmon-through-ohop-creek.

————. "Tribes Leverage Almost $9 Million in Grants to Restore Salmon Habitat (And Provide Jobs)." *Northwest Treaty Tribe News*, December 13, 2015. nwtreatytribes.org/treaty-tribes-leverage-almost-9-million-in-grants-to-restore-salmon-habitat.

Nussbaum, Ronald A., Edmund D. Brodie Jr., and Robert M. Storm. *Amphibians and Reptiles of the Pacific Northwest*. Moscow: University of Idaho Press, 1983.

Nylen, Thomas H. "Spatial and Temporal Variations of Glaciers (1913–1994) on Mount Rainier and the Relation with Climate." Master's thesis, Portland State University, 2001.

O'Connell, E. "Tagging Juvenile Fish to Track Restored Habitat Use." *Northwest Indian Fisheries Commission NWIFC News*, Spring 2013. nwifc.org/w/wp-content/uploads/downloads/2013/04/2013_1_spring_magazine.pdf.

Old-Growth Definition Task Force. *Interim Definitions for Old Growth Douglas-Fir and Mixed Conifer Forests in the Pacific Northwest and California*. U.S. Department of Agriculture. Forest Service. Pacific Northwest Research Station. Research Note PNW-447, 1986.

Oliver, Chadwick Dearing. "Managing Forest Landscapes and Sustaining Old Growth." In Spies and Duncan, eds., *Old Growth in a New World*.

Oliver, Marie. "The Ties That Bind: Pacific Northwest Truffles, Trees and Animals in Symbiosis." *PNW Science Findings* 118. Portland, OR: Pacific Northwest Research Station, USDA Forest Service, 2009.

Olympic Peninsula Audubon Society. "Barred Owl Management." 2015. olympicpeninsulaaudubon.org/conservation/barred-owl-management.

Oregon Fish and Wildlife Office. *Experimental Removal of Barred Owls to Benefit Threatened Northern Spotted Owls*. Draft Environmental Impact Statement. Portland, OR: U.S. Fish and Wildlife Service, 2012.

————. "Restoring Vitality to Our Northwest Forests." Northern Spotted Owl Information Site, 2012. www.fws.gov/oregonfwo/species/Data/NorthernSpottedOwl/main.asp.

————. *Record of Decision: Experimental Removal of Barred Owls to Benefit Threatened Spotted Owls*. Portland, OR: U.S. Fish and Wildlife Service, 2013.

————. "Barred Owl Study Update." www.fws.gov/oregonfwo/articles.cfm?id=149489616.

————. "Barred Owl Threat." www.fws.gov/oregonfwo/articles.cfm?id=149489615.

Orting School District. *Emergency Response Plan*. ortingschools.schoolwires.net/cms/lib03/WA01919463/Centricity/Domain/79/Emergency%20response%20plan.pdf.

Pacific Northwest Seismic Network. "Volcano Seismicity: Mount Rainier". pnsn.org/volcanoes/mount-rainier#description.

Parker, Geoffrey G. "Structure and Microclimate of Forest Canopies." In *Forest Canopies*, Margaret D. Lowman and Nalini M. Nadkarni, eds. San Diego: Academic Press, 1995.

Pelto, Mauri. "Paradise Glacier Ice Caves Lost." *From a Glaciers Perspective: Glacier Change in a World of Climate Change*, April 29, 2010. glacierchange.wordpress.com/2010/04/29/paradise-glacier-ice-caves-lost.

Perry, Don. "Ohop Monitoring is an Important Part of Restoration." *Yil-me-hu: The Nisqually Watershed Salmon Recovery Newsletter*, Winter 2013–2014.

———. "Ohop Valley Restoration Enters Phase III." *Yil-me-hu: The Nisqually Watershed Salmon Recovery Newsletter*, Winter 2013–2014.

———. "Restoration of the Nisqually River Estuary Is Rapidly Restoring Feeding and Growth Opportunities for Juvenile Chinook Salmon." *Yil-me-hu: The Nisqually Watershed Salmon Recovery Newsletter*, Winter 2013–2014.

———. "Spawning Surveys Critical to Salmon Recovery." *Yil-me-hu: The Nisqually Watershed Salmon Recovery Newsletter*, Winter 2013–2014.

Pierce County Department of Emergency Management. *Mount Rainier Volcanic Hazards Plan*. Working draft, 2008.

Pierce County Geographic Information Services. *Map 13, Pierce County Shoreline Master Program Update, % Impervious Surface by Basin*, 2007. www.piercecountywa.gov/DocumentCenter/View/26966/Map-13-Percentage-of-Impervious-Surface-by-Basin?bidId=.

Pierce County Public Works and Utilities Surface Water Management Division. *2013 Report Card: Surface Water Health*. www.co.pierce.wa.us/ArchiveCenter/ViewFile/Item/2612.

———. *Pierce County Rivers Flood Hazard Management Plan*, vol. I. www.co.pierce.wa.us/ArchiveCenter/ViewFile/Item/4672.

———. *2014 Report Card: Surface Water Health*. www.co.pierce.wa.us/ArchiveCenter/ViewFile/Item/3812.

———. *2015 Report Card: Surface Water Health*. www.co.pierce.wa.us/ArchiveCenter/ViewFile/Item/4672.

Pierson, Thomas C., Nathan J. Wood, and Carolyn L. Driedger. "Reducing Risk from Lahar Hazards: Concepts, Case Studies, and Roles for Scientists." *Journal of Applied Volcanology* 3:16 (2014).

Pojar, Jim, and Andy MacKinnon. *Plants of the Pacific Northwest Coast*. Vancouver: British Columbia Ministry of Forests/Lone Pine Publishing, 2004.

Poole, Alan J., editor. "Birds of North America Online." Ithaca: Cornell Laboratory of Ornithology. birdsna.org.

Porter, Stephen C. *Recent Glacier and Climate Variations in the Pacific Northwest*. Seattle: Quaternary Research Center, University of Washington, n.d.

Pringle, Patrick T. "Buried and Submerged Forests of Oregon and Washington: Time Capsules of Environmental and Geologic History," *Western Forester* 59:2 (2014), 14–15, 22. www.forestry.org/media/docs/westernforester/2014/WF_March_April_May_2014-1.pdf.

Pringle, Patrick T., Newell P. Campbell, Katherine M. Reed, and Anne C. Heinitz. *Roadside Geology of Mount Rainier National Park and Vicinity*. Olympia, WA: Washington Division of Geology and Earth Resources Information Circular 107, 2008.

Pritzker, Barry M. *Native Americans: An Encyclopedia of History, Culture and Peoples*. Vol. 1. Santa Barbara: ABC-CLIO, 1998.

Puget Sound Partnership. *State of the Sound, Report on the Puget Sound Vital Signs*. Olympia, WA. pspwa.app.box.com/2015-SOS-vitalsigns-report.

———. *State of the Sound, Report to the Community*. Olympia, WA, 2015. pspwa.app.box.com/2015-SOS-community-report.

Puyallup River Watershed Council. *Puyallup River Watershed Assessment*. Unpublished draft, 2014.

Raley, Catherine M. and Aubry, Keith B. "Foraging Ecology of Pileated Woodpeckers in Coastal Forests of Washington." *Journal of Wildlife Management* 70:5 (2007), 1266–75.

Ralph, C. John, George L. Hunt Jr., Martin G. Raphael, and John F. Piatt. *Ecology and Conservation of the Marbled Murrelet.* Gen. Tech. Rep. PSW-GTR-152. Albany, CA: Pacific Southwest Research Station, Forest Service, U.S. Department of Agriculture, 1995.

Raphael, Martin G., Gary A. Falxa, Katie M. Dugger, Beth M. Galleher, Deanna Lynch, Sherri L. Miller, S. Kim Nelson, and Richard D. Young. *Northwest Forest Plan: The First 15 Years (1994–2008): Status and Trend of Nesting Habitat for the Marbled Murrelet.* Gen. Tech. Rep. PNW-GTR-848. Portland, OR: U.S. Department of Agriculture, Forest Service, Pacific Northwest Research Station, 2011.

Ray, Verne F. *Handbook of Cowlitz Indians.* Seattle: Northwest Copy Company (1966).

———. "Native Villages and Groupings of the Columbia Basin." *Pacific Northwest Quarterly* 27 (1936), 99–152.

Reed, Richard J. "Flying Saucers over Mount Rainier." *Weatherwise*, April 1958.

Reese, Gary Fuller. *Mount Rainier National Park Place Names.* Tacoma: Tacoma Public Library, 2009.

———. *Nothing Worthy of Note Transpired Today: The Northwest Journals of August V. Kautz.* Tacoma, WA: Tacoma Public Library, 1978.

Reid, Fiona A. *A Field Guide to Mammals of North America North of Mexico*, 4th ed. New York: Houghton Mifflin (2006).

Remias, D., U. Lutz-Meindl, and C. Lutz. "Photosynthesis, Pigments and Ultrastructure of the Alpine Snow Alga *Chlamydomonas nivalis.*" *European Journal of Phycology* 40:3 (2005), 259–68.

Rice, David G. *Archaeological Test Excavations in Fryingpan Rockshelter, Mount Rainier National Park.* Washington State University Laboratory of Anthropology Report of Investigations No. 33. Pullman, WA, 1965.

———. "Indian Utilization of the Cascade Mountain Range in South Central Washington." *The Washington Archaeologist* 8:1 (1964).

Ridgway, Robert. *Color Standards and Color Nomenclature.* Baltimore: A. Hoen and Company, 1912.

Riedel, Jon, and Michael A. Larrabee. *Mount Rainier National Park Glacier Mass Balance Monitoring Annual Report, Water Year 2011: North Coast and Cascades Network.* Natural Resource Data Series NPS/NCCN/NRDS—2015/752. Fort Collins, CO: National Park Service, 2015.

Rochefort, Regina M. "The Influence of White Pine Blister Rust (*Cronartium ribicola*) on Whitebark Pine (*Pinus albicaulis*) in Mount Rainier National Park and North Cascades National Park Service Complex, Washington." *Natural Areas Journal*, 28:3 (2008), 290–98.

Rochefort, Regina M., and Darin D. Swinney. "Human Impact Surveys in Mount Rainier National Park: Past, Present, and Future." *USDA Forest Service Proceedings* RMRS-P-15-VOL-5, 2000.

Rochefort, Regina M., and David L. Peterson. "Temporal and Spatial Distribution of Trees in Subalpine Meadows of Mount Rainier National Park, Washington, USA." *Arctic and Alpine Research* 28:1 (1996), 52–59.

Rochefort, Regina M., and Stephen T. Gibbons. "Impact Monitoring and Restoration in Mount Rainier National Park. *Park Science.* 13 (1993): 29-30.

———. "Mending the Meadow: High-Altitude Restoration in Mount Rainier National Park." *Restoration and Management Notes* 10:2 (1992), 120–26.

Rochefort, Regina M., Laurie L. Kurth, Tara W. Carolin, Jon L. Riedel, Robert R. Mierendorf, Kimberly Frappier, and David L. Steensen. "Mountain Ecosystem Restoration." In *Restoring the Pacific Northwest: The Art and Science of Ecological Restoration in Cascadia.* D. Apostol and M. Sinclair, eds. Washington, DC: Island Press, 2006.

Rochefort, Regina M., Matt Albright, and Pat Milliren. "The Propagation of Three Greenhouse Programs in the Pacific Northwest." *Park Science* 19:1 (1998), 17–19.

Rodolfo, Kelvin S. "Notes from the Hazard from Lahars and Jökulhlaups." In *Encyclopedia of Volcanoes*, H. Sigurdsson et al., eds. San Diego: Academic Press, 2000.

Rook, Erik J., Dylan G. Fischer, Rebecca D. Seyferth, Justin L. Kirsch, Carri J. LeRoy, and Sarah Hamman. "Responses of Prairie Vegetation to Fire, Herbicide and Invasive Species Legacy." *Northwest Science* 85:2 (2011).

Ross, Mickey. "Old Growth: Failures of the Past and Hope for the Future." In Spies and Duncan, eds., *Old Growth in a New World*.

Ruby, Robert H. and John A. Brown. *A Guide to the Indian Tribes of the Pacific Northwest*. Norman: University of Oklahoma Press, 1986.

Rybolt, Brian Robert. "Roads Less Traveled: Access, Automobiles, and Recreation in Mount Rainier National Park's Wilderness Areas." Master's thesis, University of Washington, 2013.

Sagan, Carl. "Unidentified Flying Objects." *Encyclopedia Americana*. New York: Grolier, Inc., 1967.

Salwasser, Hal. "Regional Conservation of Old-Growth Forest in a Changing World: A Global and Temporal Perspective." In Spies and Duncan, eds., *Old Growth in a New World*.

Schmoe, Floyd W. "Indians Still Visit Park." *Mount Rainier Nature Notes* 3:15 (1926).

———. *A Year in Paradise*. Seattle: Mountaineers, 1979.

Schullery, Paul. *Island in the Sky: Pioneering Accounts of Mount Rainier, 1833–1894*. Seattle: Mountaineers, 1987.

Schultz, Cheryl B., Erica Henry, Alexa Carleton, Tyler Hicks, Rhiannon Thomas, Ann Potter, Michele Collins, Mary Linders, Cheryl Fimbel, Scott Black, Hannah E. Anderson, Grace Diehl, Sarah Hamman, Rod Gilbert, Jeff Foster, Dave Hays, David Wilderman, Roberta Davenport, Emily Steel, Nick Page, Patrick L. Lilley, Jennifer Heron, Nicole Kroeker, Conan Webb, and Brian Reader. "Conservation of Prairie-Oak Butterflies in Oregon, Washington, and British Columbia." *Northwest Science,* 85:2 (2011), 361–88.

Sellet, Frédéric, Russell Dean Greaves, and Pei-Lin Yu, eds. *Archaeology and Ethnoarchaeology of Mobility*. Gainesville: University Press of Florida, 2006.

Shaw, David C. and Ronald J. Taylor. "Pollination Ecology of an Alpine Fell-Field Community in the North Cascades." *Northwest Science* 60:1 (1986), 21–31.

Shaw, George Coombs. *The Chinook Jargon and How to Use It: A Complete and Exhaustive Lexicon of the Oldest Trade Language of the American Continent*. Seattle: Rainier Printing Co., 1909.

Shogren, Elizabeth. "To Save Threatened Owl, Another Species Is Shot." *National Public Radio*. January 16, 2014. www.npr.org/2014/01/15/262735123/to-save-threatened-owl-another-species-is-shot.

Sinclair, Marcia, Ed Alverson, Patrick Dunn, Peter Dunwiddie, and Elizabeth Gray. "Bunchgrass Prairies." In *Restoring the Pacific Northwest: the Art and Science of Ecological Restoration in Cascadia*. D. Apostol, and M. Sinclair, eds. Washington, DC: Island Press, 2006.

Sisson, T. W., J. E. Robinson, and D. D. Swinney. "Whole-Edifice Ice Volume Change A.D. 1970 to 2007/2008 at Mount Rainier, Washington, Based on LiDAR Surveying." *Geology* 39, no. 7 (2011): 639–42.

Skinner, Brian J., Stephen C. Porter, Jeffrey Park, and Harold L. Levin. *Dynamic Earth: An Introduction to Physical Geology*, 5th ed. Hoboken, NJ: Wiley Custom Services, 2008.

Slater, G. L., K. Foley, R. R. McGregor, G. Singleton, R. Hetschko, J. Lynch, and B. Altman. *An Update on Western Bluebird Recovery in the Pacific Northwest*. Poster presented at Cascadia Prairie Oak Partnership Conference. Eugene, Oregon, 2018.

Small, Ernest and Paul M. Catling. *Canadian Medicinal Crops*. Ottawa: NRC Research Press, 1999.

Smith, Allan H. *Takhoma: Ethnography of Mount Rainier National Park*. Pullman: Washington State University Press, 2006.

Smith, Marian W. *The Puyallup-Nisqually*. New York: Columbia University Press, 1940.

Spalding, Shelley A. "Bull Trout: Big Beautiful Fish in Small Pristine Streams." Master's thesis, The Evergreen State College, 1994.

Speich, Steven M., Terrence R. Wahl, and David A. Manuwal. "The Numbers of Marbled Murrelets in Washington Marine Waters." *In Status and Conservation of the Marbled Murrelet in North America*, Harry R. Carter and Michael L. Morrison, Editors. Proceedings of a symposium held at the annual meeting of the Pacific Seabird Group, 16-20 December 1987. Volume 5 Number 1, (1992), 48-60.

Spies, Thomas A. "Science of Old-Growth, Or a Journey into Wonderland." In Spies and Duncan, eds., *Old Growth in a New World*.

Spies, Thomas A., and Sally L. Duncan, eds. *Old Growth in a New World: A Pacific Northwest Icon Reexamined*. Washington, DC: Island Press, 2009.

Spies, Thomas A., Sally L. Duncan, K. Norman Johnson, Frederick J. Swanson, and Denise Lach. "Conserving Old Growth in a New World." In Spies and Duncan, eds., *Old Growth in a New World*.

Sprenger, Carson B. and Peter W. Dunwiddie. "Fire History of a Douglas-Fir-Oregon White Oak Woodland, Waldron Island, Washington." *Northwest Science*, 85:2 (2011), 108–19.

Stanley, Amanda G., Peter W. Dunwiddie, and Thomas N. Kaye. "Restoring Invaded Pacific Northwest Prairies: Management Recommendations from a Region-Wide Experiment." *Northwest Science* 85:2 (2011), 233–46.

Stein, Julie K. and Laura S. Phillips. *Vashon Island Archaeology: A View from Burton Acres Shell Midden*. Seattle: University of Washington Press, 2002.

Stewart, D. B., and Canada Department of Fisheries and Oceans. "Fish Life History and Habitat Use in the Northwest Territories: Bull Trout (*Salvelinus confluentus*)." Canadian Manuscript Report of Fisheries and Aquatic Sciences, 2801. Ottawa: Fisheries and Oceans Canada, 2007.

Stewart, Hilary. *Cedar: Tree of Life of the Northwest Coast Indians*. Seattle: University of Washington Press, 1984.

Stinson, Derek W. *Periodic Status Review for the Streaked Horned Lark in Washington*. Olympia: Washington Department of Fish and Wildlife, 2016. wdfw.wa.gov/publications/01774/wdfw01774.pdf.

Survival of American Indians Association. "As Long As the Rivers Run: One Indian Family's Struggle to Hold onto Their Treaty Rights." Video recording, 1971.

Suttles, Wayne and Barbara Lane. "Southern Coast Salish." In *Handbook of North American Indians*, W. C. Sturtevant, ed. Smithsonian Institution. Washington, DC: U.S. Government Printing Office, 1998.

Tangley, Laura. *The Buzz about Bumblebees*. National Wildlife Federation. March 31, 2014. www.nwf.org/Magazines/National-Wildlife/2014/AprilMay/Gardening/Bumblebees.

Taulman, James F. "Late Summer Activity Patterns in Hoary Marmots." *Northwestern Naturalist* 71, no. 2 (1990), 21–26.

———. "Vocalizations of the Hoary Marmot, *Marmota caligata*." *Journal of Mammalogy* 58, no. 4 (1977), 681–83. doi:10.2307/1380026.

Taylor Jr., Avery L., and Eric D. Forsman. "Recent Range Extensions of the Barred Owl in Western North America, Including the First Records for Oregon." *The Condor* 78 (1976): 560–61.

Taylor, Walter P. "A Distributional and Ecological Study of Mount Rainier, Washington." *Ecology* 3:3 (1922), 214–36.

Teit, James Alexander. "The Middle Columbia Salish." Edited by Franz Boas. *University of Washington Publications in Anthropology* 2, no. 4 (1928).

Theobald, Elli J., Ian Breckheimer, and Janneke HilleRisLambers. "Climate Drives Phenological Reassembly of a Mountain Wildflower Community." *Ecology* 98 (2017).

Thomas, Edward Harper. *Chinook: A History and Dictionary of the Northwest Coast Trade Jargon*. Portland, OR: Binford and Mort, 1935.

Thomas, Matthew A. and Paul Kennard. *Topographic and Hydrologic Insight for the Westside Road Problem: Mount Rainier National Park*. Natural Resource Report NPS/MORA/NRR-2015/1057. Fort Collins, CO: National Park Service, 2015.

Thompson, William Francis. "An Approach to Population Dynamics of the Pacific Red Salmon." *Transactions of the American Fisheries Society* 88:3 (1959), 206–9.

Thorson, R. M. "Ice-sheet Glaciation of the Puget Lowland, Washington, during the Vashon Stade (Late Pleistocene)." *Quaternary Research* 13 (1980), 303–21.

Thrush, Coll. "The Lushootseed Peoples of Puget Sound Country." University of Washington Libraries Digital Collections. Seattle: University of Washington Libraries, n.d. content.lib. washington.edu/aipnw/thrush.html.

Tolmie, William Fraser. *The Journals of William Fraser Tolmie, Physician and Fur Trader*. Vancouver, CA: Mitchell Press Limited, 1963.

Tomback, Diana F. "Clark's Nutcracker: Agent of Regeneration." In Tomback et al., *Whitebark Pine Communities*.

———. "The Impact of Seed Dispersal by Clark's Nutcracker on Whitebark Pine: Multi-Scale Perspective on a High Mountain Mutualism." In *Mountain Ecosystems: Studies in Treeline Ecology*. Gabriele Broll, and Beate Keplin. Berlin: Springer, 2005.

Tomback, Diana F., and Katherine C. Kendall. "Biodiversity Losses: The Downward Spiral." In Tomback et al. *Whitebark Pine Communities*.

Tomback, Diana F., Stephen F. Arno, and Robert E. Keane. "The Compelling Case for Management Intervention". In Tomback et al., *Whitebark Pine Communities*.

———, eds. *Whitebark Pine Communities: Ecology and Restoration*. Washington, DC: Island Press, 2001.

Tomback, Diana F., and Yan B. Linhart. "The Evolution of Bird-Dispersed Pines." *Evolutionary Ecology* 4, no. 3 (1990), 185–219.

Troutt, David. "Director's Corner." *Yil-me-hu: The Nisqually Watershed Salmon Recovery Newsletter*, Winter 2013–2014.

Tucker, Dave. *Geology Underfoot in Western Washington*. Missoula, MT: Mountain Press, 2015.

Turek, Michael F. and Robert H. Keller Jr. "Sluskin: Yakima Guide to Mount Rainier." *Columbia: The Magazine of Northwest History* 5:1 (1999).

Tweiten, Michael A. *The Interaction of Changing Patterns of Land Use, Sub-Alpine Forest Composition and Fire Regime at Buck Lake, Mount Rainier National Park, USA*. Prepared for United States Department of the Interior National Park Service: Seattle, WA, 2007.

U.S. Army Corps of Engineers, Seattle District. "Corps Awards $112 Million Contract for Nation's Largest Fish Passage Facility." News release no. 18-009. March 15, 2018. www.nws.usace.army. mil/Media/News-Releases/Article/1467005/corps-awards-112-million-contract-for-nations-largest-fish-passage-facility.

———. *Final Environmental Assessment for Mud Mountain Dam Upstream Fish Passage, Pierce County, Washington* (2015).www.nws.usace.army.mil/Portals/27/docs/environmental/ resources/2015EnvironmentalDocuments/Draft_FINAL_Fish_Trap_and_Barrier_Structure_ EA_4-30-15_FINAL.pdf?ver=2015-05-05-125719-650.

U.S. Army Environmental Command. *Army Compatible Use Buffer Program*. Unpublished, 2011.

U.S. Department of the Interior. *Correspondence Concerning Native Americans Hunting and Camping in Yakima Park at Mount Rainier*. Unpublished, 1915.

U.S. Department of the Interior: National Park Service. *The Salmon Life Cycle*. Olympic National Park, Washington, 2015. www.nps.gov/olym/learn/nature/the-salmon-life-cycle.htm.

U.S. Environmental Protection Agency Region 10 Seattle, Washington. *Fourth Five-Year Review Report for Commencement Bay Nearshore/Tideflats Superfund Site Pierce County, Washington*. Seattle: U.S. Environmental Protection Agency, 2014. www.epa.gov/region10/pdf/sites/ cb-nt/4th_fyr_cbnt_2014.pdf.

U.S. Environmental Protection Agency. *Salish Sea: Chinook Salmon*. www.epa.gov/salish-sea/chinook-salmon.

———. *Superfund*. www2.epa.gov/superfund.

U.S. Fish and Wildlife Service. *2007 Draft Recovery Plan for the Northern Spotted Owl, Strix occidentalis caurina: Merged Options 1 and 2*. Portland, OR: U.S. Fish and Wildlife Service.

———. *Biological Opinion for the Carbon River Access Management Plan*. Unpublished, 2011.

———. "Endangered and Threatened Wildlife and Plants; Determination of Endangered Status for Taylor's Checkerspot Butterfly and Threatened Status for the Streaked Horned Lark; Final Rule." *Federal Register* 78 (2013): 192.

————. *Recovery Plan for the Threatened Marbled Murrelet (Brachyramphus marmoratus) in Washington, Oregon, and California.* Portland, OR: U.S. Fish and Wildlife Service, 1997.

————. *Revised Recovery Plan for the Northern Spotted Owl (Strix occidentalis caurina).* Portland, OR: U.S. Fish and Wildlife Service, 2011.

U.S. Geological Survey. *The Cataclysmic 1991 Eruption of Mount Pinatubo, Philippines.* U.S. Geological Survey Fact Sheet 113-97 (1997). pubs.usgs.gov/fs/1997/fs113-97.

————. *Water Properties: Dissolved Oxygen.* 2015. water.usgs.gov/edu/dissolvedoxygen.html.

Urton, James. "With Climate Change, Mount Rainier Floral Communities Could 'Reassemble' with New Species Relationships, Interactions". *UW News.* University of Washington. November 17, 2017. www.washington.edu/news/2017/11/07/with-climate-change-mount-rainier-floral-communities-could-reassemble-with-new-species-relationships-interactions.

Vallance, James W. "Lahars." In *Encyclopedia of Volcanoes,* H. Sigurdsson et al., eds. San Diego: Academic Press, 2000.

————. *Notes on the Origin, Distribution and Hazards of Lahars at Mount Rainier.* Unpublished report, 2001.

Vallance, James W., and Kevin M. Scott. "The Osceola Mudflow from Mount Rainier—Sedimentology and Hazard Implications of a Huge Clay-Rich Debris Flow." *Geological Society of America Bulletin* 109:2 (1997), 143–63.

Vallance, James W., and Patrick T. Pringle. "Lahars, Tephra, and Buried Forest: The Postglacial History of Mount Rainier." In Pringle et al., *Roadside Geology of Mount Rainier National Park and Vicinity.* Washington Division of Geology and Earth Resources Information Circular, 2008.

Vallance, J. W., S. P. Schilling, G. Devoli, M. E. Reid, M. M. Howell, and D. L. Brien. *Lahar Hazards at Casita and San Cristobal Volcanoes, Nicaragua.* United States Geological Survey Open-File Report 2001-468. pubs.usgs.gov/of/2001/0468.

Vancouver, George. *A Voyage of Discovery to the North Pacific Ocean and Round the World, 1791–1795,* vol. 2. John Vancouver, ed. London: G. G. and J. Robinson, 1798.

van Rooyen, Jason C., Joshua M. Malt, and David B. Lank. "Relating Microclimate to Habitat Availability: Edge Effects on Nesting Habitat Availability for the Marbled Murrelet." *Northwest Science,* 85:4 (2011), 549–61.

Vitt, Dale H., Janet E. Marsh, and Robin B. Bovey. *Mosses, Lichens and Ferns of Northwest North America.* Edmonton, AB: Lone Pine Publishing, 1988.

Von Hagen, Bettina. "Unexplored Potential of Northwest Forests." In Spies and Duncan, eds., *Old Growth in a New World.*

Wahl, Terence R., Bill Tweit, and Steven G. Mlodinow. *Birds of Washington: Status and Distribution.* Corvallis: Oregon State University Press, 2005.

Walker, Lawrence R. *The Biology of Disturbed Habitats.* New York: Oxford University Press, 2012.

Walkup, Laura C. and Justin Ohlschlager. *Nisqually Glacier Mapping and Stagnant Ice Survey 2012 Field Protocol.* Unpublished, 2012.

Washington Department of Fish and Wildlife. *Salmon Hatcheries Overview.* 2015. wdfw.wa.gov/hatcheries/overview.html. wdfw.wa.gov/fishing/management/hatcheries.

Washington State Department of Ecology. *Floodplain Management.* 2019. www.ecy.wa.gov/programs/sea/floods/index.html.

————. *Toxics Cleanup in Commencement Bay.* 2015.

————. *Water Quality Assessment and 303(d) List.* 2019. www.ecy.wa.gov/programs/Wq/303d/index.html.

————. *Water Quality Improvement Project Commencement Bay Area: Dioxin.* 2015.

————. *Water Quality Improvement Project Puyallup River Basin Area.*

————. *Threatened and Endangered Wildlife in Washington: 2012 Annual Report.* Listing and Recovery Section, Wildlife Program, Washington Department of Fish and Wildlife, Olympia, 2013.

Waterhouse, F. Louise, Alan E. Burger, David B. Lank, Peter K. Ott, Elsie A. Krebs, and Nadine Parker. "Using the Low-Level Aerial Survey Method to Identify Marbled Murrelet Nesting Habitat." *British Columbia Journal of Ecosystems and Management* 10, no. 1 (2009): 80–96. jem-online.org/index.php/jem/article/view/413.

Welch, Paul S. "Glacier Oligochaeta from Mt. Rainier." *Transactions of the American Microscopical Society*, 35:2 (1916), 85–124.

Wellman, Candace. *Peace Weavers: Uniting the Salish Coast through Cross-Cultural Marriages*. Pullman: Washington State University Press, 2017.

Whitlock, Cathy and Margaret A. Knox. "Prehistoric Burning in the Pacific Northwest: Human Versus Climatic Influences." In *Fire, Native Peoples and the Natural Landscape*. Thomas R. Vale, ed. Washington, DC: Island Press, 2002.

Wiens, J. David, Katie M. Dugger, Krista E. Lewicki, and David C. Simon. *Effects of Experimental Removal of Barred Owls on Population Demography of Northern Spotted Owls in Washington and Oregon—2015 Progress Report*. U.S. Geological Survey Open-File Report 2016-1041.

Wikipedia. *Paradise Ice Caves*. en.wikipedia.org/wiki/Paradise_Ice_Caves.

Wild Pacific Salmon. *Wild Alaskan Chinook*. pacificwild.org.

Wilkes, Charles. *Narrative of the United States Exploring Expedition, during the Years 1838, 1839, 1840, 1841, 1842*, vol. 2. London: Ingram, Cooke, and Co., 1852.

Wilkinson, Charles. *Messages from Frank's Landing: A Story of Salmon, Treaties and the Indian Way*. Seattle: University of Washington Press, 2000.

Williams, Hill. *The Restless Northwest: A Geological Story*. Pullman: Washington State University Press, 2002.

Wilma, David and Walt Crowley. *Tacoma–Thumbnail History*. HistoryLink.org Essay 5055, January 17, 2003. www.historylink.org/File/5055.

Wilson, Anna, Kevin Bacher, Ian Breckheimer, Jessica Lundquist, Regina Rochefort, Elli Theobald, Lou Whiteaker, and Janneke HilleRisLambers. "Monitoring Wildflower Phenology Using Traditional Science, Citizen Science and Crowdsourcing Approaches." *Park Science* 33:1 (2016).

Witham, C. S. "Volcanic Disasters and Incidents—A New Database." *Journal of Volcanology and Geothermal Research*, 148 (2005): 191–233.

Wolf, Adrian Lance. "Bird Use of Epiphyte Resources in an Old-Growth Coniferous Forest of the Pacific Northwest." Master's thesis, The Evergreen State College, 2009.

Wood, Nathan J., and Christopher E. Soulard. *Community Exposure to Lahar Hazards from Mount Rainier, Washington*. Scientific Investigations Report 2009-5211. United States Geological Survey, 2009.

———. "Variations in Population Exposure and Sensitivity to Lahar Hazards from Mount Rainier." Washington. *Journal of Volcanology and Geothermal Research* 188 (2009): 367–378.

Wright, Judy. (2001). *Puyallup Tribe of Indians. Project Report: Cushman Site*. Puyallup Tribal Council, 2001.

Wright, Judy and Carol Ann Hawks. *History of the Puyallup Tribe of Indians*. Tacoma, WA: Puyallup Tribal Council, 1997.

Zouhar, Kris. "Cytisus scoparius, C. striatus." Fort Collins, CO: USDA Forest Service, Rocky Mountain Research Station, Fire Sciences Laboratory, 2005. www.fs.fed.us/database/feis/plants/shrub/cytspp/all.html.

Zweifel, Matthew K. and Connie S. Reid. "Prehistoric Use of the Central Cascade Mountains of Eastern Washington State: An Overview of Site Types and Potential Resource Utilization." *Archaeology in Washington*, vol. 3. Bellingham, WA: Association for Washington Archaeology, 1991.

Zwinger, Ann H., and Beatrice E. Willard. *Land above the Trees: A Guide to American Alpine Tundra*. New York: Harper and Row, 1972.

Index

About the Author

Photo by Doug Jones

After graduating from Western Washington University's Huxley College of the Environment with a B.S. in Environmental Education, Jeff Antonelis-Lapp began a 37-year teaching career by leading outdoor programs and teaching science. He later earned a M.Ed. in Science Education from the University of Washington.

As a faculty member at The Evergreen State College in Olympia, Jeff taught on western Washington Indian reservations for 10 years before teaching environmental education, natural history, and writing on campus. He retired as emeritus faculty in 2015. Seasonal work at Mount Rainier in the late 1970s ignited his connection to the mountain that endures today. He has climbed it, hiked all of the park's mapped trails, and completed the Wonderland Trail six times.

Jeff and his wife Valerie reside in Enumclaw, where they raised their two children.